Practicing Biology

A Student Workbook

for

Biological Science,

Fifth Edition

Scott Freeman

Jean Heitz
University of Wisconsin-Madison

Cynthia Giffen
University of Wisconsin-Madison

PEARSON

Boston   Columbus   Indianapolis   New York   San Francisco   Upper Saddle River
Amsterdam   Cape Town   Dubai   London   Madrid   Milan   Munich   Paris   Montréal   Toronto
Delhi   Mexico City   São Paulo   Sydney   Hong Kong   Seoul   Singapore   Taipei   Tokyo

Vice President/Editor-in-Chief: Beth Wilbur
Acquisitions Editor: Michael Gillespie
Project Editor: Brady Golden
Managing Editor, Production: Michael Early
Production Project Manager: Jane Brundage
Production Management and Composition: S4Carlisle Publishing Services
Executive Marketing Manager: Lauren Harp
Media Project Manager: Shannon Kong
Cover Production: Seventeenth Street Studios
Manufacturing Buyer: Jeffrey Sargent

Cover Photo Credit: Physignathus cocincinus, © Eric Isselée/Fotolia

1 2 3 4 5 6 7 8 9 10—EBM—17 16 15 14 13

PEARSON

ISBN 10: 0-321-88647-X
ISBN 13: 978-0-321-88647-7

Contents

Contents   v

Contents   vii

Contents

Contents

About *Practicing Biology: A Student Workbook*

What does Practicing Biology: A Student Workbook contain?

The activities in this workbook focus on key ideas, principles, and concepts that are basic to understanding biology. The overall organization follows that of *Biological Science*, 4th edition. Key principles or processes developed in activities are often revisited and integrated in subsequent activities. Although the individual activities may vary in the thought processes required and in their specific biological content, these are the overall goals of this workbook:

- Allow students to discover what they know and, more important, what they don't know.
- Help students to discover and modify any misconceptions in their understanding of biology.
- Provide students with opportunities to synthesize and apply what they have learned to novel situations.

What kinds of activities are included in the workbook?

The activities take several different forms:

Leading questions. In these activities, students are asked a series of leading questions that are designed to build their basic understanding of principles. Leading questions are generally followed by additional questions that give students the opportunity to apply what they have learned to new situations.

Concept mapping, diagramming, and drawing. These activities, which include drawing exercises, concept maps, and flow diagrams, are designed to help students organize information and ideas and develop an understanding of how various pieces of information are interrelated.

Modeling. Modeling activities provide students with instructions for building models of dynamic biological processes that occur at the molecular, cellular, and physiological levels. Modeling can help students both develop and test their understanding of processes that are generally invisible to the naked eye.

Process of Science. Process of science activities encourage students to practice using scientific thought processes. They are designed to give students a better understanding of how the knowledge they gain in class can be applied to

- propose experiments,
- predict possible outcomes of experiments, and
- interpret experimental data.

Reviewing. This group of activities provides an opportunity for students to review and integrate key ideas and principles in biology. The reviews are generally followed by activities that require students to apply their knowledge.

Teaching. In the teaching activities, students examine ideas, principles, and concepts from the instructor's point of view. Most instructors will agree that their understanding of a process, idea, or concept increases when they work to help someone else learn it. These activities give students opportunities to develop deeper understanding by "putting themselves in the instructor's shoes."

Name _____   Course/Section _____

Date _____   Professor/TA _____

  Activity 2.1 A Quick Review of Elements and Compounds

1. Figure 2.3 (page 20 of *Biological Science*, 5th ed.) lists the chemical elements that
    occur naturally in the human body. Similar percentages of these elements are found
    in most living organisms.

a. In what abiotic (nonlife) chemical forms are these elements often found in nature?	b. In what chemical form(s) do animals need to obtain these elements?	c. In what chemical form(s) do plants need to obtain these elements?

2. A chemical element cannot be broken down to other forms by chemical reactions.
    Each element has a specific number of protons, neutrons, and electrons.

    a. What is the name of the following element, and how many protons, neutrons, and
        electrons does it have?

Name	Number of protons	Number of neutrons	Number of electrons

b. What information do you need to calculate or to determine the following?

The atomic number of an element	The mass number of an element	The weight in atomic mass units (daltons) of one atom of an element

c. What are the atomic number, mass number, and weight in atomic mass units (daltons) of the element shown in part a?

Atomic number	Mass number	Weight in atomic mass units

3. One mole of an element or compound contains 6.02×10^{23} atoms or molecules of the element or compound. One mole of an element or compound has a mass equal to its mass number (or molecular weight) in grams. For example, 1 mole of hydrogen gas (H_2) contains 6.02×10^{23} molecules and weighs 2 g.

a. What is the weight of 1 mole of pure sodium (Na)?	b. How many molecules of Na are in 1 mole of Na?

c. How would you determine how many grams are in a mole of any chemical element or compound?

4. One atom of Na can combine with one atom of Cl (chlorine) to produce one molecule of NaCl (table salt).

a. If Cl has 17 electrons, 17 protons, and 18 neutrons, what is its mass number?	b. What is the mass number of NaCl?	c. How many grams of NaCl equal a mole of NaCl?

d. If you wanted to combine equal numbers of Na and Cl atoms in a flask, how much Cl would you have to add if you added 23 g of Na? (Explain the reasoning behind your answer.)

e. To make a one-molar (1 *M*) solution of NaCl, you need to add 1 mole of NaCl to distilled water to make a final volume of 1 L (1000 ml). A 1 *M* solution is said to have a molarity of 1. If you added 2 moles of NaCl to distilled water to make a final volume of 1 L, you would make a 2 *M* solution and its molarity would equal 2.

You make up a 1 *M* solution of NaCl.

How many molecules of NaCl are in the 1 *M* NaCl solution?	How many molecules of NaCl are there per ml of the solution?

f. Next, you divide this 1 *M* solution of NaCl into four separate flasks, putting 250 ml into each flask.

How many grams of NaCl are in each flask?	How many molecules of NaCl are in each flask?	How many molecules of NaCl are there per ml of distilled water?	What is the molarity of NaCl in each of the four flasks?

5. The summary formula for photosynthesis is

$$6\,CO_2 + 6\,H_2O \longrightarrow C_6H_{12}O_6 + 6\,O_2$$

a. How many molecules of carbon dioxide and water would a plant have to use to produce 3 molecules of glucose ($C_6H_{12}O_6$)?	b. How many moles of carbon dioxide and water would a plant have to use to produce 2 moles of glucose?

c. Refer to the summary formula for photosynthesis. If you know the number of molecules or moles of any of the reactants used (or products produced), how would you calculate the number of molecules or moles of all of the other reactants needed and products produced?

6. A biologist places a plant in a closed chamber. A sensor in the chamber maintains the carbon dioxide level at the normal atmospheric concentration of 0.03%. Another sensor allows the biologist to measure the amount of oxygen produced by the plant over time. If the plant produces 0.001 mole of oxygen in an hour, how much carbon dioxide had to be added to the chamber during that hour to maintain the atmospheric concentration of 0.03%?

7. Refer to pages 19–30 of your textbook for a description of types of chemical bonds: nonpolar and polar covalent bonds, ionic bonds, and hydrogen bonds. (The molecule diagrammed here can also be represented by the formula CH_3COOH.)

$$
\begin{array}{c}
\quad\; H \\
\quad\; | \\
H-C-C \overset{\displaystyle =O}{\underset{\displaystyle \diagdown OH}{}} \\
\quad\; | \\
\quad\; H
\end{array}
$$

Explain how you could determine which of the bonds between elements in this molecule are polar or nonpolar covalent bonds, ionic bonds, and hydrogen bonds.

8. O_2 and NH_3 are both small covalent molecules found in cells. NH_3 is extremely soluble in the aqueous environment of the cell, whereas O_2 is relatively insoluble. What is the basis for this difference in solubility between the two molecules? In reaching your answer, draw the structures of the molecules as valence shell diagrams (as in Figure 2.8, page 23). Given these diagrams, consider the types of interactions each molecule could have with water.

Activity 2.1   5

Name _____   Course/Section _____

Date _____   Professor/TA _____

  Activity 2.2 A Quick Review of the Properties of Water

1. Compounds that have the capacity to form hydrogen bonds with water are said to be hydrophilic (water loving). Those without this capacity are hydrophobic (water fearing).

$$H-\underset{\underset{H}{\vert}}{\overset{\overset{H}{\vert}}{C}}-C\underset{\diagdown\; OH}{\overset{\diagup\!\!\diagup\, O}{}}$$	Is the molecule on the left hydrophilic or hydrophobic? Explain your answer.

2. In addition to being polar, water molecules can dissociate into hydronium ions (H_3O^+, often described simply as H^+) and hydroxide ions (OH^-). The concentration of each of these ions in pure water is 10^{-7}. Another way to say this is that the concentration of hydronium ions, or H^+ ions, is one out of every 10 million molecules. Similarly, the concentration of OH^- ions is one in 10 million molecules.

   a. The H^+ ion concentration of a solution can be represented as its pH value. The pH of a solution is defined as the negative $\log_{10}$ of the hydrogen ion concentration. What is the pH of pure water?

   b. Refer to the diagram of the molecule of acetic acid in question 1. The –COOH group can ionize to release an H^+ ion into solution. If you add acetic acid to water and raise the concentration of H^+ ions to 10^{-4}, what is the pH of this solution?

3. Life as we know it could not exist without water. All the chemical reactions of life occur in aqueous solution. Water molecules are polar and are capable of forming hydrogen bonds with other polar or charged molecules. As a result, water has the following properties:

    A. H_2O molecules are cohesive; they form hydrogen bonds with each other.
    B. H_2O molecules are adhesive; they form hydrogen bonds with polar surfaces.
    C. Water is a liquid at normal physiological (or body) temperatures.
    D. Water has a high specific heat.
    E. Water has a high heat of vaporization.
    F. Water's greatest density occurs at 4°C.

Explain how these properties of water are related to the phenomena described in parts a–h below. More than one property may be used to explain a given phenomenon.

a. During the winter, air temperatures in the northern United States can remain below 0°C for months; however, the fish and other animals living in the lakes survive.

b. Many substances—for example, salt (NaCl) and sucrose—dissolve quickly in water.

c. When you pour water into a 25-ml graduated cylinder, a meniscus forms at the top of the water column.

d. Sweating and the evaporation of sweat from the body surface help reduce a human's body temperature.

e. A bottle contains a liquid mixture of equal parts water and mineral oil. You shake the bottle vigorously and then set it on the table. Although the law of entropy favors maximum randomness, this mixture separates into layers of oil over water.

f. Water drops that fall on a surface tend to form rounded drops or beads.

g. Water drops that fall on your car tend to bead or round up more after you polish (or wax) the car than before you polished it.

h. If you touch the edge of a paper towel to a drop of colored water, the water will move up into (or be absorbed by) the towel.

Name _____   Course/Section _____

Date _____   Professor/TA _____

Extend Your
Understanding   **Activity 2.3 Are silicon-based life forms possible?**

Silicon is very similar to carbon in its chemical properties. For example, both are in the same group of the periodic table, and both Si and C have 4 electrons in their outer shells. Based on these similarities, some science fiction writers have suggested that life-forms on other planets could be silicon-based. Given what you know about the two elements, their reactivity or electronegativity, and the types of compounds they form, indicate whether it is possible that life could be Si-based as well as C-based. Feel free to use additional resources to make an argument for or against a world where silicon forms the building blocks of life.

Name _____   Course/Section _____

Date _____   Professor/TA _____

 Activity 3.1 What properties do proteins have?

1. Functional groups can modify the properties of organic molecules. In the table below, indicate whether each functional group is polar or nonpolar and hydrophobic or hydrophilic. Which of these functional groups are found in proteins? Use Table 2.3 on page 37 and Figure 3.2 on page 43 of *Biological Science*, 5th edition, to fill in the table.

Functional group	Polar or nonpolar?	Hydrophobic or hydrophilic?	Found in all proteins?	Found in many proteins?
–OH				
–CH$_2$				
–COOH				
–NH$_2$				
–SH				
–PO$_4$				

2. Twenty amino acids are commonly utilized in the synthesis of proteins. These amino acids differ in the chemical properties of their side chains (also called R-groups). What properties does each of the following R-groups have? (*Note:* A side chain may display more than one of these properties.)

R-group	Basic, acidic, or neutral	Polar or nonpolar	Hydrophilic or hydrophobic
a. CH_2 — CH — CH_3 CH_3			
b. CH_2 — C($-O$)$=O$			
c. CH_2 — CH_2 — CH_2 — CH_2 — NH_3^+			
d. CH_2 — OH			

3. Polypeptides and proteins are made up of linear sequences of amino acids. In its functional form, each protein has a specific three-dimensional structure or shape. Interactions among the individual amino acids and their side chains play a major role in determining this shape.

   a. How are amino acids linked together to form polypeptides or proteins? What is this type of bond called?

b. Define the four structures of a protein.	c. What kinds of bonds hold each of these structures together?
Primary:	
Secondary:	
Tertiary:	
Quaternary:	

d. On the figures below, label at least one of each of the following characteristics of proteins: α-helix, β-pleated sheet, H-bonding along peptide-bonded backbone, and van der Waals interaction.

(a) Cro protein, a dimer

(b) Hemoglobin, a tetramer

α_1 α_2

β_1 β_2

e. Describe how the structures above might change if denatured by high heat or by a chemical that can denature protein.

4. Use your understanding of the proteins in living organisms to predict the outcome of the following experiments. Be sure to explain your reasoning.

Experiment a: A globular protein that is ordinarily found in aqueous solution has these amino acids in its primary structure: glutamic acid, lysine, leucine, and tryptophan. Predict where you would find each amino acid: in the interior portion of the protein (away from water) or on the outside of the protein (facing water). (Refer to Figure 3.10 on page 49 of the textbook.)

Experiment b: Drawn below is part of the tertiary structure of a protein showing the positions of two amino acids (aspartic acid and lysine). Replacing lysine with another amino acid in the protein may change the shape and function of the protein. Replacing lysine with which type(s) of amino acid(s) would lead to the least amount of change in the tertiary structure of this protein? (Refer to Figure 3.11a on page 50 of the textbook.)

$$-CH_2-CH_2-CH_2-CH_2 \boxed{ NH_3^+ \quad {}^-O } \overset{\displaystyle O}{\overset{\displaystyle \|}{C}}-CH_2-$$

Ionic bond

Lysine

Aspartic acid

Name _____   Course/Section _____

Date _____   Professor/TA _____

Activity 3.2 What factors affect chemical reactions in cells?

Construct a concept map of general metabolism using the terms in the list below,
according to the following guidelines:

- Keep in mind that there are many ways to construct a concept map.
- Begin by writing each term on a separate Post-it note or piece of paper.
- Organize the terms into a map that indicates how the terms are associated or
  related.
- Draw lines between terms, and add action phrases to the lines to indicate how the
  terms are related.
- If you are doing this activity in small groups in class, explain your map to
  another group when you finish it.

Here is an example:

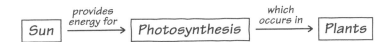

Terms

peptide bonds	activation energy	allosteric regulation
proteins	ΔG / free energy	activator
α-helix	endergonic	four-step enzyme-mediated reaction sequence or metabolic pathway (A⟶B⟶C⟶D)
primary structure	exergonic	
secondary structure	enzymes	
tertiary structure	catalysts	intermediate compound
β-pleated sheet	competitive inhibitor	
R-groups	noncompetitive inhibitor	end-product
hydrogen bonds		feedback inhibition
substrate or reactant (ligand)	active site	
	product	

Use the understanding you gained from doing the concept map to answer the following questions.

1. Reduced organic compounds tend to contain stored energy in C–H bonds. As a general rule, the greater the number of C–H bonds, the greater the amount of potential energy stored in the molecule. Answer each question in the chart as it relates to the two reactions shown at the top. Be sure to explain the reasoning behind your answers.

	Reaction 1: $CH_4 + 2\,O_2 \rightarrow 2\,H_2O + CO_2$ (methane)	Reaction 2: $6\,CO_2 + 6\,H_2O \rightarrow C_6H_{12}O_6 + 6\,O_2$
a. Is the reaction exergonic or endergonic?		
b. Is the reaction spontaneous?		
c. Is the reaction anabolic or catabolic?		
d. Is ΔG (the change in free energy) positive or negative?		

2. All metabolic reactions in living organisms are enzyme mediated. Each enzyme is specific for one (or only a very few similar types of) reaction. Given this, there are approximately as many different kinds of enzymes as there are reactions.

   a. What characteristics do all enzymes share?

   b. What characteristics can differ among enzymes?

3. How can enzyme function be mediated or modified? To answer, complete columns a and b below.

a. What factors can modify enzyme function?	b. What effect(s) can each of these factors have on enzyme function?

   c. What role(s) can modification of enzyme function play in the cell?

Name _____   Course/Section _____

Date _____   Professor/TA _____

Activity 3.3 How can changes in experimental conditions affect enzyme-mediated reactions?

1. You set up a series of experiments to monitor the rates of a reaction. The reaction is an enzyme-mediated reaction in which $A \longrightarrow B + C$. For each experiment in this series, you continuously add the reactant A and monitor its concentration so that the amount of A remains constant over time.

   For each group of experiments, explain how the differences in experimental conditions could affect the reaction.

   a. You compare two side-by-side experiments. In experiment 1, you use x amount of the enzyme. In experiment 2, you use $2x$ amount of the same enzyme.

   b. You compare two side-by-side experiments. In both you use equal amounts of the enzyme. In experiment 3, you allow the products to accumulate over time. In experiment 4, you remove the products from the system as they are produced.

   c. In the next two experiments, you use equal amounts of the enzyme. You run experiment 5 at 20°C and experiment 6 at 25°C.

   d. In two final experiments, you use equal amounts of the enzyme. You run experiment 7 at pH 6 and experiment 8 at pH 8.

2. Enzyme function can be inhibited or regulated by the presence of chemicals that mimic either the reactants or the products.

   a. How do competitive and noncompetitive inhibition of an enzyme differ?

   b. What are allosteric enzymes? What function(s) can they serve in reaction sequences?

3. An enzyme catalyzes the reaction X $\longrightarrow$ Y + Z. In a series of experiments, it was found that substance A inhibits the enzyme.

   - When the concentration of X is high and A is low, the reaction proceeds rapidly and Y and Z are formed.
   - As the concentration of A increases, the reaction slows, regardless of whether X is present in high or low concentration.
   - If the concentration of A is high (relative to X), the reaction stops.
   - If the concentration of A again decreases, the reaction will ultimately resume.

   What type of enzyme regulation is described here? Explain or justify your answer.

4. In an enzymatic pathway, A, B, and C are intermediates required to make D, and 1, 2, and 3 are enzymes that catalyze the designated reactions:

$$A \xrightarrow{\ 1\ } B \xrightarrow{\ 2\ } C \xrightarrow{\ 3\ } D$$
$$\downarrow$$
$$E$$

This is analogous to what happens in a factory. In a leather goods factory, for example, the leather (A) is cut into the parts needed for shoes (B). The shoe parts are sewn together (making C), and C is packaged for shipping as D. Now shoe sales are dropping and backpack sales (E) are increasing. As a result, the manager of the factory decides to switch production from shoes to backpacks.

a. Where should the shoe-making process be shut down—at step 1, 2, or 3? Explain.

b. In a cell, if an excess of a chemical product D arises, where should this synthetic pathway be shut down in the cell? Explain your reasoning.

c. What type(s) of enzyme regulation is/are most likely to occur in the cell in this type of feedback system? Explain your reasoning.

Name _____   Course/Section _____

Date _____   Professor/TA _____

 Activity 4.1 What makes a compound a nucleic acid?

1. Closely related macromolecules often have many characteristics in common. For example, they share many of the same chemical elements and functional groups. Therefore, to separate or distinguish closely related macromolecules, you need to determine how they differ and then target or label that difference.

   a. What makes RNA different from DNA?

   b. If you wanted to use a radioactive or fluorescent tag to label only the RNA in a cell and not the DNA, what compound(s) could you label that is/are specific for RNA?

   c. If you wanted to label only the DNA, what compound(s) could you label?

2. You want to use a radioactive tracer that will label only the protein in an RNA virus. Assume the virus is composed of only a protein coat and an RNA core. Which of the following would you use? Be sure to explain your answer.

   a. radioactive P   b. radioactive N   c. radioactive S   d. radioactive C

3. Watson and Crick used modeling to determine the structure of DNA. What information led them to determine the following?

   a. DNA is double stranded.

   b. The two strands of nucleotides in DNA run antiparallel.

   c. The nucleotides in the two antiparallel strands are arranged such that a purine is always paired with a pyrimidine.

4. For each nucleic acid strand shown below, indicate whether it is DNA or RNA, label each of the bases, and write out the nucleotide sequence, starting at the 5′ end. For assistance, refer to Figure 4.1 on page 58 in *Biological Science*, 5th edition.

a. Type of nucleic acid: _____ Nucleotide sequence as A, G, C, etc.: 5′ ___ ___ ___ ___ 3′	b. Type of nucleic acid: _____ Nucleotide sequence as A, G, C, etc.: 5′ ___ ___ ___ ___ 3′

5. Many scientists believe that RNA molecules were the first macromolecules that emerged to allow some of the fundamental processes of life to be conducted. Indicate which of the following statements supports the RNA world hypothesis, and be sure to justify your answer.

   a. RNA includes the base uracil (U) instead of thymidine (T).

   b. DNA is more stable than RNA.

   c. Unlike DNA, RNA can catalyze chemical reactions.

   d. Because RNA uses ribose instead of deoxyribose, RNA can be used as an energy source.

Name _____   Course/Section _____

Date _____   Professor/TA _____

 Activity 5.1 How does structure affect function in carbohydrates?

1. Polysaccharides, complex carbohydrates, have a wide range of structures that affect how these macromolecules can function in organisms. Fill out the chart below to help you understand the characteristics of different polysaccharide structures and the roles they play in living organisms. (Refer to Summary Table 5.1 on page 77 of *Biological Science*, 5th edition.)

	Structural components	3-D structural arrangement	Role(s) the polysaccharide play(s) in organisms	Type(s) of organisms in which compound is found
Starch				
Glycogen				
Cellulose				

Chitin			
Peptidoglycan			

2. Compare the structures of starch and glycogen. Compare the structures of cellulose, chitin, and peptidoglycan. How are each of these suited to their functions?

Activity 5.1

Name _____   Course/Section _____

Date _____   Professor/TA _____

 Activity 6.1 What roles do lipids play in organisms?

1. Fill out the chart below to help you understand the characteristics of different types of lipids and the roles they play in living organisms.

	Molecular components	Molecular arrangement	Role(s) the lipid plays in an organism
Steroid			
Phospholipid			
Fat			

2. Lipids as a group are defined as being hydrophobic, or insoluble in water. As a result, this group includes a fairly wide range of compounds—for example, fats, oils, waxes, and steroids such as cholesterol.

    a. How are fatty acids and glycerol linked together to form fats (triglycerides)?

    b. How do phospholipids differ from triglycerides?

    c. How are the structures of steroids, phospholipids, and fats suited to their different functions?

3. You have a friend who lost 20 pounds of fat on a diet. Where did the fat go (how was it lost)? Be sure to explain your reasoning.

Name _____   Course/Section _____

Date _____   Professor/TA _____

Activity 6.2 What controls the movement of materials into and out of the cell?

1. To be alive, most cells must maintain a relatively constant internal environment. To do this, they must be able to control the movement of materials into and out of the cell.

   What characteristics of the cell membrane determine what gets into the cell and what doesn't? That is, what determines the permeability of a cell or organelle membrane?

   To answer these questions, first consider the answers to the following questions:

a. If a cell membrane were composed of only a phospholipid bilayer, what properties would it have?	b. What different roles or functions do membrane proteins serve?	c. Why are some cell types more permeable to a substance (for example, sodium ions) than others?

   Using your understanding of the answers in a–c, now answer these questions: What characteristics of the cell membrane determine what gets into the cell and what doesn't? That is, what determines the permeability of a cell or organelle membrane?

2. You design an experiment to test the effect(s) various compounds have on the osmotic potential of a model cell. You know that substances dissolved in aqueous or gaseous solutions tend to diffuse from regions of higher concentration to regions of lower concentration.

You fill each of three (20-ml) dialysis bags half full with one of these substances:

- 5% by weight of glucose in distilled water
- 5% by weight of egg albumin (protein) in distilled water
- 5% by weight of glass bead (one glass bead) in distilled water

The dialysis bag is permeable to water but impermeable to glucose, albumin, and glass bead.

a. If the final weight of each bag is 10 g, how many grams of glucose, albumin, and glass bead were added to each bag?

b. The molecular weight of the protein is about 45,000 atomic mass units, and the molecular weight of glucose is about 180 atomic mass units. How can you estimate the number of molecules of glucose in the 5% solution compared to the number of albumin molecules in the 5% solution?

c. You put the dialysis bags into three separate flasks of distilled water. After 2 hours, you remove the bags and record these weights:

Dialysis bag	Weight
Glucose	13.2 g
Albumin	10.1 g
Glass bead	10.0 g

How do you explain these results? (*Hint:* Consider the surface-area-to-volume ratio of each of the three substances and review pages 26 and 27 of *Biological Science,* 5th edition.)

d. What results would you predict if you set up a similar experiment but used 5% glucose and 5% sucrose?

Name _____   Course/Section _____

Date _____   Professor/TA _____

 ## Activity 6.3 How is the structure of a cell membrane related to its function?

Membranes compartmentalize the different functions of living cells. The cell membrane is a barrier between the cell or organism and its environment. Similarly, within the cell, membranes of organelles separate the different reactions of metabolism from each other. Use the supplies provided in class or devise your own at home to develop a model of a cell membrane. Developing models of systems can help you understand their overall structure as well as their function(s).

Building the Model

- Include in the membrane the phospholipid bilayer (phosphate heads and fatty-acid tails) as well as the integral proteins.
- Design integral proteins that serve the functions of facilitated diffusion and active transport.
- Indicate how the various types of integral proteins might differ in structure and operation.

Use the understanding you gain from your model to answer the questions on the next pages.

1. Substances can move across the membrane via simple diffusion, facilitated diffusion, or active transport.

	a. Where does it occur in the membrane?	b. Does it require transport protein?	c. Does it require the input of energy?
Simple diffusion			
Facilitated diffusion			
Active transport			

d. What functions might each of the three types of diffusion serve in an independent cell such as a *Paramecium* or an amoeba?

e. What functions might each of the three types of diffusion serve in a multicellular organism—for example, a human or a tree?

2. What would you need to observe or measure to determine whether a substance was moved across a membrane via each type of diffusion?

Simple diffusion	Facilitated diffusion	Active transport

3. The ratios of saturated to unsaturated phospholipids in an organism's membranes can change in response to changes in environmental conditions.

a. How do the properties of a membrane that contains a low percentage of unsaturated phospholipids compare with those of a membrane that contains a high percentage of unsaturated phospholipids?

b. Considering what you know about the properties of saturated and unsaturated fatty acids, would you expect an amoeba that lives in a pond in a cold northern climate to have a higher or lower percentage of saturated fatty acids in its membranes during the summer as compared to the winter? Explain your answer.

4. A fish is removed from a contaminated lake. You determine that a particular toxin is present in its cells at concentration A = 1500 µg/L. You place the fish in a tank full of clean water (A = 0 µg/L), and several days later you measure the toxin concentration in the cells (level B).

a. On the graphs below, predict how the toxin concentrations in the fish and in the water will change over time if:
(i) The toxin is water soluble.
(ii) The toxin is fat soluble.

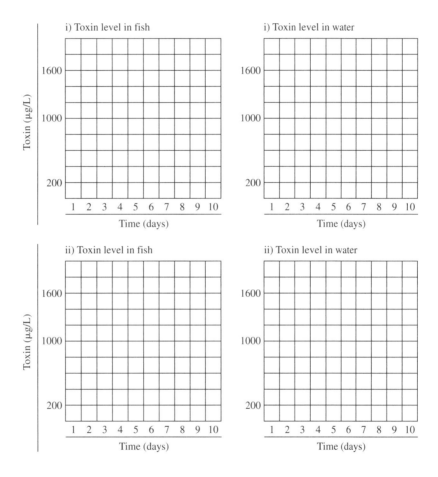

b. After making your hypothesis, you test it by measuring the toxin levels in the fish at various times during its several days in the tank. You observe that the level of toxin drops in the fish from 1500 µg/L to 750 µg/L and then stabilizes at 750 µg/L. You test the water in the tank and find that after it stabilizes, toxin also is present in the water at concentration 750 µg/L.

- Which of your predictions fit these data?
- Which of the following processes is most likely eliminating the toxin from the fish?
  (i)   passive transport
  (ii) first active, then passive transport
  (iii) first passive, then active transport
  (iv) active transport

c. Given the situation in part b, what should you do, in the short term, to continue to reduce the toxin level in the fish below 750 µg/L?

Name _____   Course/Section _____

Date _____   Professor/TA _____

Extend Your
Understanding **Activity 6.4 Effects of Missing or Nonfunctional Ion**
                 Channels in Cell Membranes

- Cystic fibrosis transmembrane conductance regulator (CFTR) proteins function in cell membranes to allow chloride ions across cell membranes. Individuals with cystic fibrosis (CF) have abnormal CFTR proteins that do not allow Cl^- to move across cell membranes. Chloride channels are essential to maintain osmotic balance inside cells. Without properly functioning Cl^- channels, water builds up inside the cell. One result is a thickening of mucus in the lungs and air passages.

You are doing research on a disease, and you hypothesize that there is a defect in an ion channel of the cell membrane, similar to the defect in the CFTR protein in individuals with CF. Diagram or model the production of a normal membrane ion channel. Based on your understanding of ion channel functions, propose at least three different alterations that could result in a nonfunctional or missing ion channel. What questions would you need to answer to determine which of these may be correct?

Name _____   Course/Section _____

Date _____   Professor/TA _____

Extend Your
Understanding **Activity 6.5 What effect does cholesterol have on phospholipid membrane permeability?**

Figure 6.10 on page 91 of your textbook shows the results of an experiment designed to test whether cholesterol within a phospholipid bilayer has an effect on membrane permeability. If the results were different, as shown in the graph below, how would the conclusion change? What additional questions would you ask, or what experiments would you propose, to better understand the relationship between membrane composition, permeability to glycerol, and temperature?

Name _____   Course/Section _____

Date _____   Professor/TA _____

Integrate Your
Understanding **Chapters 3–6 What do carbohydrates, lipids, proteins, and
nucleic acids have in common? How are they different?**

Refer to the figure (Some Simple Chemistry) on the next page when doing this activity.

**Part A. Answer the questions. Then use your answers to develop simple rules for
identifying carbohydrates, lipids, proteins, and nucleic acids.**

1. What is the approximate C:H:O ratio in each of the following types of
   macromolecules?

Carbohydrates	Lipids	Proteins	Nucleic acids

2. Which of the compounds listed in question 1 can often be composed of C, H, and O
   alone?

3. Which of the compounds can be identified by looking at the C:H:O ratios alone?

Some Simple Chemistry

Compound	Basic components	➡ Reaction ➡	Product

Carbohydrates:

Sugars, starches, glycogen, cellulose

6C hexose   [dehydration reaction]   Disaccharide   +1 H₂O

Lipids:

Fats, oils, waxes, cholesterol

Glycerol + 3 fatty acids ➡ Triglyceride or fat   +3 H₂O
[dehydration reaction]

Proteins:

Enzymes, structural proteins

Amino acid   (amino group) (carboxyl group)   [dehydration reaction]   Dipeptide   (peptide bond)   +1 H₂O

Nucleic acids:

DNA, RNA

(PO₄) + Ribose + Base (Base = A, U, G, or C) ➡ RNA (etc.)

(PO₄) + Deoxyribose + Base (Base = A, T, G, or C) ➡ DNA (etc.) (hydrogen bonds)

4. What other elements are commonly associated with each of these four types of macromolecules?

	Carbohydrates	Lipids	Proteins	Nucleic acids
Always contain P				
Generally contain no P				
Always contain N				
Generally contain no N				
Frequently contain S				
Generally contain no S				

5. Based on your answers to questions 1–4 and what you learned from Chapters 3–6 in *Biological Science,* 5th edition, what simple rule(s) can you use to identify the following macromolecules?

Macromolecule	How to identify
Nucleic acids	
DNA versus RNA	
Proteins	
Carbohydrates	
Lipids	

Part B. Carbohydrate, lipid, protein, or nucleic acid? Name that structure!

Based on the rules you developed in Part A, identify the following compounds as carbohydrates, lipids, amino acids, polypeptides, or nucleic acids. In addition, indicate whether each is likely to be polar or nonpolar, and hydrophilic or hydrophobic.

1) $C_{17}H_{35}COOH$ +

$$\begin{array}{c} H \\ | \\ H-O-C-H \\ | \\ H-O-C-H \\ | \\ H-O-C-H \end{array} \longrightarrow \begin{array}{c} H \\ | \\ C_{17}H_{35}COO-C-H \\ | \\ C_{17}H_{35}COO-C-H \\ | \\ C_{17}H_{35}COO-C-H \end{array}$$

2)

$$HC = C - CH_2 \overset{H}{\underset{NH_2}{C}} - COOH$$

$$H^+\!-\!N \underset{\underset{H}{C}}{\overset{N}{\diagdown}} \overset{N}{\diagup}_{H}$$

3)

$$\underset{HO}{\overset{O}{\diagdown}}C - \overset{H}{\underset{H}{C}} - \overset{}{\underset{H}{N}} - \overset{O}{\underset{}{C}} - \overset{OH}{\underset{\underset{H}{C}}{CHCH_3}} - \overset{}{\underset{H}{N}} - \overset{O}{\underset{}{C}} - \overset{CH_2}{\underset{\underset{H}{C}}{}} - \overset{H}{\underset{H}{N}}$$

4) $NH_2 - \overset{\overset{\displaystyle NH_2}{||}}{C} - NH - CH_2 - CH_2 - CH_2 - CH - \overset{O}{\underset{\underset{\underset{H}{N-H}}{}}{C}} \overset{}{\diagdown}_{OH}$

5)

$$\begin{array}{cccc} CH_2OH & CH_2OH & CH_2OH & CH_2OH \end{array}$$

O — O, — O, — O, — OH

OH   OH   OH   OH

6) $H - \overset{\overset{\displaystyle H}{|}}{C} - \overset{\overset{\displaystyle H}{|}}{C} - \overset{\overset{\displaystyle H}{|}}{C} - \overset{\overset{\displaystyle H}{|}}{C} - \overset{H}{\underset{O}{C}}$

$\qquad OH \ OH \ OH \ OH$

Integrate Your Understanding Chapters 3–6

7)

8)

9) $H_3C-CH_2-CH_2-CH_2-CH_2-CH_2-CH_2-CH_2-CH_2-C\overset{\displaystyle O}{\underset{\displaystyle OH}{}}$

10)

Part C. Thought Questions

1. Given your understanding of energy storage in organic macromolecules, explain why carbohydrates and proteins have 4 calories/gram, but fats have 9 calories/gram, as shown on the nutrition information of all food packaging.

2. While on a safari in the jungle, your friend was bitten by a large spider. He nearly died from hemolysis, or breakage of many of his red blood cells. You analyzed the spider venom and found three enzymes: phospholipase, which degrades phospholipids; neuraminidase, which removes cell-surface carbohydrates; and protease, which degrades proteins. Which of these enzymes do you think was responsible for his near-fatal red blood cell hemolysis? Why?

Part D. Process of Science

A student, Mary, is given four samples and told they are lysine (an amino acid), lactose (a disaccharide), insulin (a protein hormone), and RNA. The samples are in test tubes marked 1, 2, 3 and 4, but Mary doesn't know which compound is in which tube. She is instructed to identify the contents of each tube.

a. In her first test, she tries to hydrolyze a portion of the contents of each tube. Hydrolysis occurs in all tubes, except tube 3.

b. In Mary's next test, she finds that tubes 1, 2, and 3 are positive for nitrogen, but only tube 2 gives a positive result when tested for the presence of sulfur.

c. The last test Mary performs shows that the compound in tube 1 contains a high percentage of phosphate.

Based on these data, fill in the following table and explain your answers.

Tube number	Contents	Explanation
1		
2		
3		
4		

Integrate Your Understanding Chapters 3–6

Use your understanding of the chemical characteristics of the four major types of macromolecules in living organisms to predict the outcome of the following experiments. Be sure to explain your reasoning.

Part D (*continued*). Experiment a: You stir 10 g of glucose and 10 ml of phospholipids in a 500-ml beaker that contains 200 ml of distilled water. Draw a diagram to show where and how the glucose and phospholipids would be distributed after you let the mixture settle for about 30 minutes.

Experiment b: You do Experiment a again, but this time you stir 10 g of glucose and 10 ml of phospholipids in a different 500-ml beaker that contains 200 ml of distilled water and 100 ml of oil. Draw a diagram to show where and how the glucose, phospholipids, and oil would be distributed after you let the solution settle for about 30 minutes.

Experiment c: To completely fill a sealed 500-ml glass container that contains 490 ml of distilled water, you inject 10 ml of phospholipids into it. (A small gasket allows the air to leave as you inject the phospholipids.) You shake this mixture vigorously and then let it settle for an hour or more. Draw a diagram to show how the phospholipids would be distributed in the container.

Name _____   Course/Section _____

Date _____   Professor/TA _____

Activity 7.1 What makes a cell a living organism?

1. Single-celled organisms and individual cells within multicellular organisms can vary greatly in appearance as well as in the functions they perform. Nonetheless, each of these cells is alive and therefore must have some common characteristics.

a. At a minimum, what structures or components must a cell contain to be alive?	b. What is the function of each structure or component listed in part a?

   c. If you consider the types of single-celled organisms that exist today, which, if any, have a structure similar to your description in part a?

2. What would you need to add to or change about the cell you described in question 1 to make it:

a. A eukaryotic animal cell?	b. A eukaryotic plant cell?

3. To get an idea of the different sizes of various cellular components, do the following calculations: Assume that the cell, its nucleus, and a globular protein—for example, an enzyme—are spherical. In addition, assume the diameter of the protein is 5 nm, the diameter of the cell is 100 μm (micrometers), and the diameter of the nucleus is 40 μm.

If you draw the globular protein as a sphere with a diameter of 2 cm (approximately the diameter of a U.S. penny), what size would each of the following measurements of the cell be if drawn to the same scale (5 nm real length = 2 cm)?	
a. The radius of a microtubule (Refer to Summary Table 7.3 in *Biological Science*, 5th ed.)	
b. The diameter of the nucleus	
c. The diameter of the cell	
d. The volume ($V = 4/3 \, P\pi r^3$) of the protein 1 nanometer cubed $(1 \, nm^3) = 1.0 \times 10^{-21}$ centimeters cubed (cm^3)	
e. The volume of the nucleus	
f. The volume of the cell	

g. The volume of the Empire State Building is $1.05 \times 10^6 \, m^3$. How many of your scaled nuclei could fit into the Empire State Building? How many of your scaled cells could fit?

h. Do the results of these calculations help you to understand how so much can be going on inside a cell at once? Explain.

Name _____   Course/Section _____

Date _____   Professor/TA _____

Extend Your
Understanding **Activity 8.1 Do chemical reaction rates increase with**
increased temperature and concentration?

Figure 8.4 on page 140 in the textbook shows data from an experiment designed to determine if chemical reaction rates increase with increased temperature and concentration. Students at Parklawn College tested the effects of temperature and concentration on reaction rate separately, and their results are shown in Figure 8.4. You are a lab technician at a biotech company, and need to determine the fastest way to make large quantities of HSO_4^-. How would you design an experiment to test the combined effects of temperature and concentration on reaction rate in order to optimize the production process? Include the following components in your experimental design: controls, replicates, and concentration and temperatures of reactants. Clearly state your predicted and null hypotheses, and draw a graph of your expected results.

Integrate Your
Understanding **Activity 9.1 A Quick Review of Energy**
Transformations

Review pages 30–32 of Chapter 2 and pages 161–165 of Chapter 9 in *Biological*
Science, **5th edition. Then complete the following discussion by either filling in the**
blank or circling the appropriate term.

To maintain life, organisms must be able to convert energy from one form to another. For
example, in the process of photosynthesis, algae, plants, and photosynthetic prokaryotes
use the energy from sunlight to convert carbon dioxide and water to glucose and oxygen
(a waste product).

The summary reaction for photosynthesis can be written as

$$6\,CO_2 + 6\,H_2O \longrightarrow C_6H_{12}O_6 + 6\,O_2$$

This type of reaction is an oxidation–reduction (or redox) reaction. This reaction is also
[**anabolic/catabolic**] and [**endergonic/exergonic**].

In redox reactions, _____ (and associated H^+ ions) are transferred from one
compound or element to another. If one compound or element loses _____ and
becomes oxidized, another must gain _____ and become reduced. For example,
in photosynthesis, water becomes [**oxidized/reduced**] (to O_2) and the _____ (and
associated H^+ ions) it "loses" in the process [**oxidize/reduce**] CO_2 to glucose.

[**Anabolic/Catabolic**] reactions "build" more complex molecules from simpler ones. To
do this they require energy input. Reactions that require the input of energy are termed
[**endergonic/exergonic**] reactions.

The reactions involved in aerobic respiration are also redox reactions:

$$C_6H_{12}O_6 + 6\,O_2 \longrightarrow 6\,CO_2 + 6\,H_2O$$

In this set of reactions, however, more complex molecules are "broken down" into
simpler ones. Glucose is broken down or becomes [**oxidized/reduced**] (to CO_2), and the
oxygen becomes [**oxidized/reduced**] (to water).

[**Anabolic/Catabolic**] reactions break down more complex molecules into simpler ones and in the process release energy. Reactions that release energy that can be used to do work are [**endergonic/exergonic**]. Therefore, aerobic respiration is a(n) [**anabolic/catabolic**] process and is [**endergonic/exergonic**].

[**Endergonic/Exergonic**] reactions are also said to be spontaneous reactions. Does this mean that if we don't keep glucose in tightly sealed containers it will spontaneously interact with atmospheric oxygen and turn into carbon dioxide and water? The answer is obviously no.

Spontaneous reactions rarely occur "spontaneously" because all chemical reactions, even those that release energy, require some addition of energy—the energy of activation—before they can occur. One way of supplying this energy is to add heat. An example is heating a marshmallow over a flame or campfire. When enough heat is added to reach (or overcome) the activation energy, the sugar in the marshmallow reacts by oxidizing. (Burning is a form of oxidation.) The marshmallow will continue to burn even if you remove it from the campfire. As the marshmallow burns, carbon dioxide and water are formed as products of the reaction, and the energy that was stored in the bonds of the sugar is released as heat.

If our cells used heat to overcome activation energies in metabolism, they would probably burn up like the marshmallow did. Instead, living systems use protein catalysts or enzymes to lower the energy of activation without adding heat. In addition, the metabolic breakdown of sugars is carried out in a controlled series of reactions. At each step or reaction in the sequence, a small amount of the total energy is released. Some of this energy is still lost as heat. The rest is converted to other forms that can be used in the cell to drive or fuel coupled endergonic reactions or to make ATP.

Activity 9.1

Name _____   Course/Section _____

Date _____   Professor/TA _____

 Activity 9.2 Modeling Cellular Respiration: How can cells convert the energy in glucose to ATP?

Using your textbook, lecture notes, and the materials available in class (or those you devise at home), model both fermentation (an anaerobic process) and cellular respiration (an aerobic process) as they occur in a plant or animal cell. Each model should be a dynamic (working or active) representation of the events that occur in glycolysis.

Building the Model

- Use chalk on a tabletop, or a marker on a large sheet of paper, to draw the cell membrane and the mitochondrial membranes.
- Use playdough or cut-out pieces of paper to represent the molecules, ions, and membrane transporters or pumps.
- Use the pieces you assembled to model the processes of fermentation and aerobic respiration. Develop a dynamic (claymation-type) model that allows you to manipulate or move glucose and its breakdown products through the various steps of both fermentation and aerobic respiration.
- When you feel you have developed a good working model, demonstrate and explain it to another student or to your instructor.

Be sure your model of **fermentation** includes and explains the actions and roles of the following:

glycolysis   ADP
cytoplasm P_i
electrons   ATP
protons   pyruvate
glucose   ethyl alcohol (or lactic acid)
NAD^+ substrate-level phosphorylation
NADH

Be sure your model of **cellular respiration** includes and explains the actions and roles of the following:

glucose	electron transport chain
oxygen	mitochondria
carbon dioxide	inner mitochondrial membrane
pyruvate	outer mitochondrial membrane
acetyl CoA	H^+
NAD^+	electrons (e^-)
NADH	chemiosmosis
FAD	ATP synthase (proton pumps)
$FADH_2$	cristae
ADP	proton gradients
P_i	oxidative phosphorylation
ATP	substrate-level phosphorylation
water	oxidative phosphorylation

Use your models and the information in Chapter 9 of *Biological Science*, 5th edition, to answer the following questions.

1. The summary formula for cellular respiration is

$$C_6H_{12}O_6 + 6\ O_2 \longrightarrow 6\ CO_2 + 6\ H_2O + Energy$$

a. Where is each of the reactants used in the overall process?		b. Where is each of the products produced in the overall process?		
$C_6H_{12}O_6$ + $6\ O_2$	$\longrightarrow$	$6\ CO_2$ +	$6\ H_2O$ +	Energy

2. In cellular respiration, the oxidation of glucose is carried out in a controlled series of reactions. At each step or reaction in the sequence, a small amount of the total energy is released. Some of this energy is lost as heat. The rest is converted to other forms that can be used by the cell to drive or fuel coupled endergonic reactions or to make ATP.

a. What is/are the overall function(s) of glycolysis?	b. What is/are the overall function(s) of the Krebs cycle?	c. What is/are the overall function(s) of oxidative phosphorylation?

3. Are the compounds listed here *used* or *produced* in:

	Glycolysis?	The Krebs cycle?	Oxidative phosphorylation?
Glucose			
O_2			
CO_2			
H_2O			
ATP			
$ADP + P_i$			
NADH			
NAD^+			

4. The cell's supply of ADP, P$_i$, and NAD$^+$ is finite (limited). What happens to cellular respiration when all of the cell's NAD$^+$ has been converted to NADH?

5. If the Krebs cycle does not require oxygen, why does cellular respiration stop after glycolysis when no oxygen is present?

6. Many organisms can withstand periods of oxygen debt (anaerobic conditions). Yeast undergoing oxygen debt converts pyruvic acid to ethanol and carbon dioxide. Animals undergoing oxygen debt convert pyruvic acid to lactic acid. Pyruvic acid is fairly nontoxic in even high concentrations. Both ethanol and lactic acid are toxic in even moderate concentrations. Explain why this conversion occurs in organisms.

7. How efficient is fermentation? How efficient is cellular respiration? Remember that efficiency is the amount of useful energy (as ATP) gained during the process divided by the total amount of energy available in glucose. Use 686 kcal as the total energy available in 1 mol of glucose and 8 kcal as the energy available in 1 mol of ATP.

Efficiency of fermentation	Efficiency of cellular respiration

8. a. Why can't cells store large quantities of ATP? (*Hint:* Consider both the chemical stability of the molecule and the cell's osmotic potential.)

   b. Given that cells can't store ATP for long periods of time, how do they store energy?

   c. What are the advantages of storing energy in these alternate forms?

9. To make a 5 *M* solution of hydrochloric acid, we add 400 ml of 12.5 *M* hydrochloric acid to 600 ml of distilled water. Before adding the acid, however, we place the flask containing the distilled water into the sink because this solution can heat up so rapidly that the flask breaks. How is this reaction similar to what happens in chemiosmosis? How is it different?

Similarities	Differences

Extend Your
Understanding **Activity 9.3 Cell Respiration—Experimental Scenarios**

a. If it takes 1000 g of glucose to grow 10 g of an anaerobic bacterium, how many grams of glucose would it take to grow 10 g of that same bacterium if it was respiring aerobically? Estimate your answer. For example, if it takes X amount of glucose to grow 10 g of anaerobic bacteria, what factor would you have to multiply or divide X by to grow 10 g of the same bacteria aerobically? Explain how you arrived at your answer.

b. Mitochondria isolated from liver cells can be used to study the rate of electron transport in response to a variety of chemicals. The rate of electron transport is measured as the rate of disappearance of O_2 from the solution using an oxygen-sensitive electrode.

    How can we justify using the disappearance of oxygen from the solution as a measure of electron transport?

c. Humans oxidize glucose in the presence of oxygen. For each mole of glucose oxidized, about 686 kcal of energy is released. This is true whether the mole of glucose is oxidized in human cells or burned in the air. A calorie is the amount of energy required to raise the temperature of 1 g of water by 1°C; 686 kcal = 686,000 cal. The average human requires about 2000 kcal of energy per day, which is equivalent to about 3 mol of glucose per day. Given this, why don't humans spontaneously combust?

d. A gene has recently been identified that encodes for a protein that increases longevity in mice. To function in increasing longevity, this gene requires a high ratio of NAD$^+$/NADH. Researchers have used this as evidence in support of a "caloric restriction" hypothesis for longevity—that a decrease in total calorie intake increases longevity. How does the requirement for a high NAD$^+$/NADH ratio support the caloric restriction hypothesis?

e. An active college-age athlete can burn more than 3000 kcal/day in exercise.
*   If conversion of one mole of ATP to ADP + P$_i$ releases about 7.3 kcal, roughly speaking, how many moles of ATP need to be produced per day in order for this energy need to be met?
*   If the molecular weight of ATP is 573, how much would the required ATP weigh in kilograms?
*   Explain these results.

Name _____   Course/Section _____

Date _____   Professor/TA _____

Activity 10.1 Modeling Photosynthesis: How can cells use the sun's energy to convert carbon dioxide and water into glucose?

Activity 10.1 is designed to help you understand:

1. The roles photosystems I and II and the Calvin cycle play in photosynthesis

2. How and why C_4 and CAM photosynthesis differ from C_3 photosynthesis

Using your textbook, lecture notes, and the materials available in class (or those you devise at home), model photosynthesis as it occurs in a plant cell. Your model should be a dynamic (working or active) representation of the events that occur in the various phases of C_3 photosynthesis.

Building the Model

- Use chalk on a tabletop, or a marker on a large sheet of paper, to draw the cell membrane and the chloroplast membranes.
- Use playdough or cut-out pieces of paper to represent the molecules, ions, and membrane transporters or pumps.
- Use the pieces you assembled to model the processes involved in C_3 photosynthesis. Develop a dynamic (claymation-type) model that allows you to manipulate or move carbon dioxide and water and its breakdown products through the various steps of the process.
- When you feel you have developed a good working model, demonstrate and explain it to another student or to your instructor.

Your model of C_3 photosynthesis should include what occurs in photosystems I and II and in the Calvin cycle. For **photosystems I and II**, be sure your model includes and explains the roles of the following:

$NADP^+$	H^+
NADPH	e^-
ADP	chemiosmosis
P_i	ATP synthase
ATP	e^- carriers in thylakoid
water and oxygen	membranes

Also indicate where in the plant cell each item is required or produced.

For the **Calvin cycle**, be sure your model includes and explains the roles of the following:

glucose

C_3 or three-carbon sugars

carbon dioxide

NADPH

ATP

Also indicate where in the plant cell each item is required or produced.

After you've modeled C_3 photosynthesis, indicate how the system would be altered for C_4 and CAM photosynthesis.

- Indicate where in the cells of the leaf PEP carboxylase exists and how it reacts to capture CO_2. Be sure to indicate the fate of the captured CO_2.
- Do the same for PEP carboxylase in CAM plants.

Use your model and the information in Chapter 10 of *Biological Science*, 5th edition, to answer the following questions.

1. The various reactions in photosynthesis are spatially segregated from each other within the chloroplast. Draw a simplified diagram of a chloroplast and include these parts: outer membrane, grana, thylakoid, lumen, stroma/matrix.

a. Where in the chloroplast do the light reactions occur?	
b. Where in the chloroplast is the chemiosmotic gradient developed?	
c. Where in the chloroplast does the Calvin cycle occur?	

2. In photosynthesis, the reduction of carbon dioxide to form glucose is carried out in a controlled series of reactions. In general, each step or reaction in the sequence requires the input of energy. The Sun is the ultimate source of this energy.

a. What is/are the overall function(s) of photosystem I?	b. What is/are the overall function(s) of photosystem II?	c. What is/are the overall function(s) of the Calvin cycle?

3. Are the compounds listed here *used* or *produced* in:	Photosystem I?	Photosystem II?	The Calvin cycle?
Glucose			
O_2			
CO_2			
H_2O			
ATP			
ADP + P_i			
NADPH			
$NADP^+$			

4. Which light reaction system (cyclic or noncyclic) would a chloroplast use in each situation?

a. Plenty of light is available, but the cell contains little NADP$^+$.	b. There is plenty of light, and the cell contains a high concentration of NADP$^+$.

5. All living organisms require a constant supply of ATP to maintain life. If no light is available, how can a plant make ATP?

Name _____   Course/Section _____

Date _____   Professor/TA _____

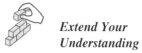

Extend Your
Understanding **Activity 10.2 Chloroplast Function under Different Conditions**

Chloroplast thylakoids can be isolated and purified for biochemical experiments. Shown below is an experiment in which pH was measured in a suspension of isolated thylakoids before and after light illumination (first arrow). At the time indicated by the second arrow, a chemical compound was added to the thylakoids. Examine these data and address the following questions.

a. Based on your understanding of the function of the chloroplasts, why does turning on the light cause the pH in the solution outside the thylakoids to increase?

b. Given the response, the chemical added was probably an inhibitor of:
    (A) oxidative phosphorylation
    (B) ATP synthase
    (C) NADPH breakdown
    (D) electron transport chain between photosystems I and II
    (E) rubisco

Justify your answer.

Name _____   Course/Section _____

Date _____   Professor/TA _____

 ## Activity 10.3 How do C$_3$, C$_4$, and CAM photosynthesis compare?

1. Carbon dioxide enters plant leaves through the stomata, and oxygen (the photosynthetic waste product) and water from the leaves exit through the stomata. Plants must constantly balance both water loss and energy gain (as photosynthesis). This has led to the evolution of various modifications of C$_3$ photosynthesis.

	C$_3$	C$_4$	CAM
Draw simplified diagrams of the cross sections of a leaf from a C$_3$, a C$_4$, and a CAM plant.			
a. How are the leaves similar?			
b. How are the leaves different?			
c. How and when does carbon dioxide get into each leaf?			
d. Which enzyme(s) (1) capture carbon dioxide and (2) carry it to the Calvin cycle?	(1) (2)	(1) (2)	(1) (2)

e. What makes C$_4$ photosynthesis more efficient than C$_3$ photosynthesis in tropical climates?

f. How is CAM photosynthesis advantageous in desert climates?

2. Photosynthesis evolved very early in Earth's history. Central to the evolution of photosynthesis was the evolution of the enzyme rubisco (an abbreviation for ribulose bisphosphate carboxylase oxidase). To the best of our knowledge, all photosynthetic plants use rubisco. Rubisco's function is to supply carbon dioxide to the Calvin cycle; however, it does this only if the ratio of carbon dioxide to oxygen is relatively high. (For comparison, a relatively high ratio of carbon dioxide to oxygen is 0.03% carbon dioxide to 20% oxygen.) When the ratio of carbon dioxide to oxygen becomes low, the role of rubisco switches and it catalyzes photorespiration, the breakdown of glucose to carbon dioxide and water.

a. Why could we call photorespiration a "mistake" in the functioning of the cell?

b. Rubisco is thought to have evolved when Earth had a reducing atmosphere. How does that hypothesis help explain this "mistake"?

Name _____   Course/Section _____

Date _____   Professor/TA _____

Activity 11.1 How are chemical signals translated into cellular responses?

Chapter 11 in *Biological Science*, 5th edition, describes at least four types of signal receptors. Three of these—G-protein-linked receptors, tyrosine-kinase receptors, and ion-channel receptors—are plasma membrane proteins. Protein receptors found in the cytoplasm, or nucleus, of the cell are the fourth type. Some signals (for example, a protein hormone) interact with signal receptors in the cell membrane to initiate the process of signal transduction. This often involves changes in a series of different relay molecules in a signal transduction pathway. Ultimately, the transduced signal initiates an intracellular response. Other types of signals (for example, steroid hormones) can diffuse through the cell membrane and interact with intracellular receptors. For example, testosterone interacts with its receptor in the cell's cytoplasm, enters the nucleus, and causes the transcription of specific genes.

To help you understand how signal transduction occurs in cells, develop dynamic (claymation-type) models of both a G-protein receptor system and a tyrosine-kinase receptor system. Use playdough or cut-out pieces of paper to represent all of the structural components and molecules listed here under each system.

G-protein receptor system	Tyrosine-kinase receptor system
signal protein	signal protein
G-protein-linked receptor	tyrosine-kinase receptor
plasma membrane	plasma membrane
inactive and active G protein	inactive and active relay proteins
GTP and GDP	ATP and ADP
inactive and active enzyme	signal transduction pathway
signal transduction pathway	

Use your models to show how signal reception by each of the systems can lead to the release of Ca^{2+} from the endoplasmic reticulum. Demonstrate and explain your models to another student group or to your instructor.

Use your models and the information in Chapter 8 of your textbook to answer the following questions.

1. How are these two systems similar? Consider both structural similarities and similarities in how the systems function.

2. How are the two systems different? Consider both structural differences and differences in how the systems function.

3. Both systems can generate elaborate multistep signal transduction pathways. These pathways can greatly amplify the cell's response to a signal; the more steps in the pathway, the greater the amplification of the signal. Explain how this amplification can occur. (Review Figure 11.13 on page 211 in your textbook.)

Name_____   Course/Section_____

Date_____   Professor/TA_____

 Activity 12.1 What is mitosis?

What is mitosis?

1. What is the overall purpose of mitosis?

2. In what types of organism(s) and cells does mitosis occur? What type of cell division occurs in bacteria?

What are the stages of mitosis?

3. The fruit fly, *Drosophila melanogaster*, has a total of eight chromosomes (four pairs) in each of its somatic cells. Somatic cells are all cells of the body except those that will divide to form the gametes (ova or sperm). Review the events that occur in the various stages of mitosis.

Keep in mind that the stages of cell division were first recognized from an examination of fixed slides of tissues undergoing division. On fixed slides, cells are captured or frozen at particular points in the division cycle. Using these static slides, early microscopists identified specific arrangements or patterns of chromosomes that occurred at various stages of the cycle and gave these stages names (interphase, prophase, and so on). Later work using time-lapse photography made it clear that mitosis is a continuous process. Once division begins, the chromosomes move fluidly from one phase to the next.

Assume you are a microscopist viewing fruit-fly cells that are undergoing mitosis. In each of the circles (cell membranes) on the following page, draw what you would expect to see if you were looking at a cell in the stage of mitosis indicated. If no circle is present, draw what you would expect to see at the given stage.

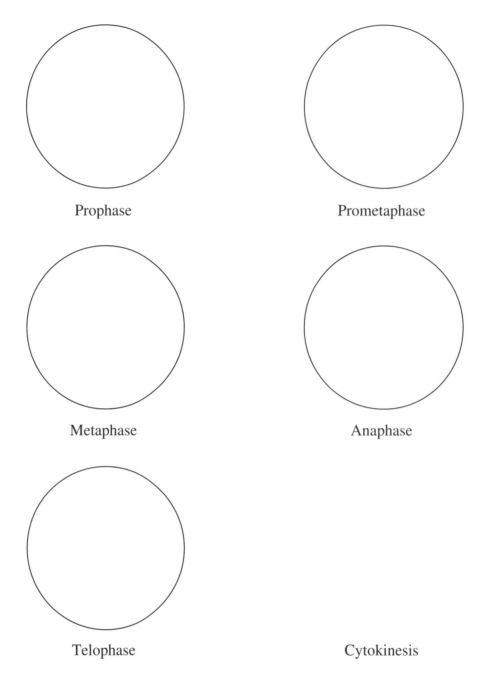

Prophase

Prometaphase

Metaphase

Anaphase

Telophase

Cytokinesis

Daughter cells in interphase

What are the products of mitosis?

4. How many cells are produced at the end of a single mitotic division?

5. How many different kinds of cells are produced at the end of a single mitotic division?

6. Six centromeres are observed in a prophase cell from another species of insect.

a. How many pairs of chromosomes does this organism contain?		
b. For each stage of mitosis, indicate the number of centromeres you would expect to find and the number of copies of chromosomes attached to each centromere.		
Stage of mitosis	Number of centromeres visible per cell	Number of chromosome copies attached to each centromere
Prophase		
Anaphase		

What controls mitosis?

7. Checkpoints in the normal cell cycle prevent cells from going through division if problems occur—for example, if the DNA is damaged.

    a. What forms do the checkpoints take? That is, how do they control whether or not cell division occurs?

b. On this page, develop a handout or diagram to explain how these checkpoints work under normal conditions. Your handout should include a description of each checkpoint, where it acts in the cell cycle, and what each checkpoint does to control cell division.

c. Cancer results from uncontrolled cell division. Explain how mutations in one or more of the checkpoints might lead to cancer.

Name _____   Course/Section _____

Date _____   Professor/TA _____

 Activity 13.1 What is meiosis?

What is meiosis?

1. What is the overall purpose of meiosis?

2. In what types of organism(s) does meiosis occur? What type of cell division occurs in bacteria?

What are the stages of meiosis?

3. The fruit fly, *Drosophila melanogaster*, has a total of eight chromosomes (four pairs) in each of its somatic cells. Somatic cells are all cells of the body except those that will divide to form the gametes (ova or sperm). Review the events that occur in the various stages of meiosis.

Keep in mind that the stages of cell division were first recognized from an examination of fixed slides of tissues undergoing division. On fixed slides, cells are captured or frozen at particular points in the division cycle. Using these static slides, early microscopists identified specific arrangements or patterns of chromosomes that occurred at various stages of the cycle and gave these stages names (interphase, prophase I, and so on). Later work using time-lapse photography made it clear that meiosis is a continuous process. Once division begins, the chromosomes move fluidly from one phase to the next.

Assume you are a microscopist viewing fruit-fly cells that are undergoing meiosis. In each of the circles (cell membranes) on the following pages, draw what you would expect to see if you were looking at a cell in the stage of meiosis indicated. If no circle is present, draw what you would expect to see at the given stage.

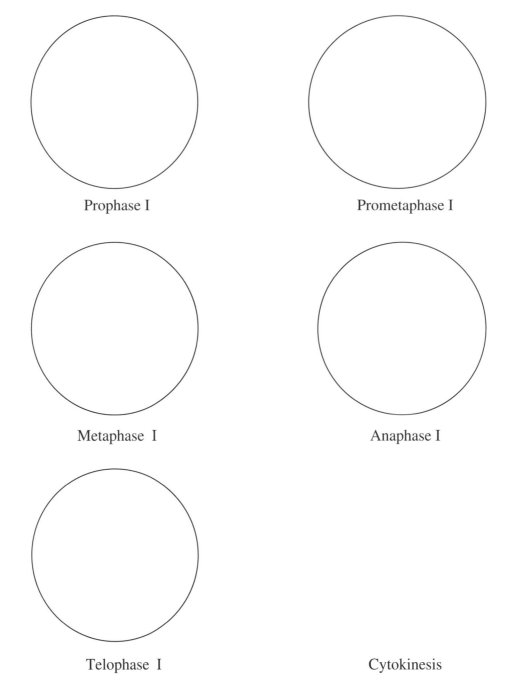

Prophase I

Prometaphase I

Metaphase I

Anaphase I

Telophase I

Cytokinesis

Follow one daughter cell through meiosis II.

Prophase II

Prometaphase II

Metaphase II

Anaphase II

Telophase II

Cytokinesis

What are the products of meiosis?

4. Consider a single cell going through meiosis.

   a. How many cells are produced at the end of meiosis?

   b. How many chromosomes, and which chromosomes, does each of the daughter cells contain?

5. Six centromeres are observed in a prophase I cell from another species of insect.

a. How many pairs of chromosomes does this organism contain?		
b. For each stage of meiosis, indicate the number of centromeres you would expect to find and the number of copies of chromosomes attached to each centromere.		
Stage of meiosis	Number of centromeres visible per cell	Number of chromosome copies attached to each centromere
Anaphase I		
Prophase II		

Extend Your
Understanding   **Activity 13.2 Sex Chromosome Trisomy due to**
                   Nondisjunction

Questions 1 to 4. Nondisjunction of sex chromosomes during human gamete formation may lead to individuals with sex chromosome trisomy. Decide whether each of the following is true or false. An individual with the sex chromosome trisomy of XXY may have resulted from nondisjunction occurring in:

T/F   1.   meiosis I in the father's sperm production

T/F   2.   meiosis II in the father's sperm production

T/F   3.   meiosis I in the mother's egg production

T/F   4.   meiosis II in the mother's egg production

Extend Your
Understanding **Activity 13.3 An Experiment Designed to Test the
"Changing-Environment Hypothesis"**

You are employed as a consultant at a large paper company, because the company is
concerned about defoliation caused by insects in its poplar plantations. Insect damage can
have severe effects on tree health, and can even cause tree mortality in large areas of forest.
The paper company is considering altering the way it grows trees from a one-species
plantation to more of a natural multispecies forest setting, in response to the insect damage.
After consulting with a forest entomologist, you find out that the insect most prevalent
in the poplar plantations is the poplar tentmaker (*Icthyura inclusa*), but you also find
the cottonwood leaf beetle (*Chrysomela scripta*) present. Most trees can reproduce both
vegetatively (asexual clones) and sexually; poplar trees are known to produce huge tracts
of clonal shoots, and a single tree can produce hundreds of thousands of seeds per year.

Given that there can be both sexual and asexual reproduction in poplar trees, you decide
that it would be a good idea to test the changing-environment hypothesis for this site,
in comparison to a nearby natural forest that has a mixture of poplar, red oak, and black
cherry. You collect data at ten 10-m^2 plots at each site (plantation and forest) based on
two factors: percentage of infected trees and percentage of trees that are clonal versus
sexually reproduced. The data you collect are provided in the chart below.

a. Fill in the missing values in the data table, and draw a graph that best describes
    the mean and variability of the data. The formula for standard error = standard
    deviation/square root of the number of samples.

	Plantation site		Forest site	
	Percent infected trees	Percent clonal	Percent infected trees	Percent clonal
Plot 1	22	84	24	45
Plot 2	19	81	23	52
Plot 3	26	76	28	55
Plot 4	24	74	17	47
Plot 5	28	82	25	56
Plot 6	18	83	16	43
Plot 7	22	79	21	49
Plot 8	25	73	26	52
Plot 9	23	86	19	51

(Continued)

	Plantation site		Forest site	
	Percent infected trees	Percent clonal	Percent infected trees	Percent clonal
Plot 10	19	84	22	48
Mean	22.6			
SD (Standard Deviation)	3.27	4.52	3.90	4.18
N (Number of samples)	10			
SE (Standard Error)	1.03			

b. What do the results from the table and graph mean? Do they support or refute the changing-environment hypothesis? What would you recommend to the paper company?

c. Propose an alternate hypothesis to explain the results in your graph.

Name _____   Course/Section _____

Date _____   Professor/TA _____

 Integrate Your
Understanding **Chapters 12–13—How do mitosis and meiosis differ?**

Review the processes of mitosis and meiosis in Chapters 12 and 13 of *Biological Science*, 5th edition, and then fill in the following chart.

1. What events occur during each phase of mitosis and meiosis?

	Interphase	Prophase	Metaphase	Anaphase	Telophase and cytokinesis
Mitosis	For example: G_1—cell growth S—DNA duplication G_2—cell growth		For example: Duplicated chromosomes, each with two sister chromatids, line up independently on the metaphase plate.		
Meiosis I					
Meiosis II					

2. Fill in this chart to summarize the major similarities and differences in the two types of cell division (mitosis vs. meiosis). For similarities, include the event(s) that always happen(s) in prophase, for example, no matter which of the cell division cycles you're describing.

	Interphase	Prophase	Metaphase	Anaphase	Telophase
a. What similarities do you see?					
b. What differences do you see?					

Amount of DNA in:	Interphase	Prophase	Metaphase	Anaphase	Telophase
Mitosis					
Meiosis I					
Meiosis II					

c If the amount of DNA in a somatic cell equals C during G_1 of interphase, then how much DNA is present in the cell during each phase of mitosis and meiosis?

d. How do the similarities in prophase of mitosis and meiosis compare with the similarities in telophase of mitosis and meiosis?

e. At what stage(s) does/do most of the differences among mitosis, meiosis I, and meiosis II occur? Why do these differences exist?

Name _____   Course/Section _____

Date _____   Professor/TA _____

 Activity 14.1 A Genetics Vocabulary Review

Mendel did not know anything about chromosomes, genes, or DNA. Because modern genetics uses vocabulary that assumes students today understand these ideas, however, it's helpful to review some key terms.

Match each commonly used genetics term with its appropriate definition or example.

Terms	Definitions and Examples
___ heterozygous	a. Blue-eyed blonde mates with brown-eyed brunette.
___ homozygous	b. *BB* or *bb*
___ monohybrid cross	c. not on sex chromosomes
___ autosomal	d. blue or brown eyes
___ genotype	e. *Bb*
___ phenotype	f. locus on a chromosome that codes for a given polypeptide*
___ gene	g. Blonde mates with brunette.
___ allele	h. *BB*, *Bb*, or *bb*
___ dihybrid cross	i. Males have only one for each gene on the X chromosome.

*Note: It is true that a gene can code for a polypeptide. However, not all genes code for polypeptides. Some code for mRNAs that produce polypeptides, but others code for other forms of RNA—for example, rRNA and tRNA.

Name _____   Course/Section _____

Date _____   Professor/TA _____

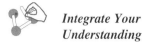 *Integrate Your*
Understanding **Chapters 13–14—Modeling Meiosis: How can diploid organisms produce haploid gametes?**

Integrate your understanding of meiosis (Chapter 13) and of basic Mendelian principles (Chapter 14) to develop a dynamic model of meiosis. When you've completed the model, use it to explain what aspects of meiosis account for Mendel's laws of segregation and independent assortment.

Building the Model

Working in groups of three or four, construct a dynamic (claymation-type) model of meiosis for the organism described on the next page. You may use the materials provided in class or devise your own.

When developing and explaining your model, be sure to include definitions or descriptions of all these terms and structures:

diploid	dominant allele
2n/n	genotype
chromosome	maternal
chromatid	paternal
chromatin	spindle
centromere (kinetochore)	spindle fibers
autosome	nuclear membrane
sex chromosome	nucleolus
sex cell	phenotype
autosome	heterozygous
crossing over	homozygous
synapsis	law of segregation
recessive allele	law of independent assortment

What genetic and chromosomal traits does your organism have?

1. Your individual is male/female (choose one). Females are XX and males are XY. For simplicity, assume that the individual is diploid with $2n = 6$, including the sex chromosomes. On one pair of autosomes (the nonsex chromosomes), the individual is heterozygous for hair color (B = brown and dominant, b = blonde and recessive). On another pair of autosomes, the organism is heterozygous for hair structure (C = curly and dominant, c = straight and recessive). Assume further that the individual's mother was homozygous dominant for both traits, and the father was homozygous recessive for both.

   a. Is your individual's hair curly or straight? Brown or blonde?

   b. What did the individual's mother's hair look like? What did the father's hair look like?

   c. What chromosomes and alleles were in the egg and the sperm that gave rise to your individual?

What does the nucleus contain?

First develop a model of a cell from your individual.

- Use chalk on a tabletop, or a marker on a large sheet of paper, to draw a cell's membrane and its nuclear membrane. The nucleus should be at least 9 inches in diameter.
- Use playdough or cut-out pieces of paper to represent your individual's chromosomes. Indicate the placement of genes on the chromosomes. Put all the chromosomes from your individual into the nucleus.
- Make a key for your model that indicates how alleles are designated and which of the chromosomes are maternal versus paternal contributions.

Then develop a model of the meiotically active cell.

- Make an identical copy of the original cell. This will be the "active" cell—that is, the one that undergoes meiosis.
- Using the "active" cell only, develop a dynamic model of meiosis. To do this, actively move the chromosomes of this one cell through a complete round of

meiosis in a sex cell. (Sex cells are the cells of the body that give rise to gametes: ova or sperm.)

- Use your model to demonstrate meiosis to another student group or to your instructor. Then use your model to answer the following questions.

What are the products of meiosis?

2. From a single sex cell going through meiosis, how many daughter cells are produced?

3. How many different kinds of gametes are produced from a single cell undergoing meiosis? (Assume no crossing over occurs.)

4. Your individual is heterozygous for two genes on separate pairs of homologous chromosomes. His or her genotype is *CcBb*. Given this information alone, how many different kinds of gametes could this individual produce? (Again assume no crossing over occurs.)

5. Compare your answer to question 4 with your answer to question 3. How do the numbers of different gametes in your answers compare? Explain any difference.

6. What aspect(s) of meiosis account(s) for Mendel's laws of:

   a. Segregation?

   b. Independent assortment?

Name _____   Course/Section _____

Date _____   Professor/TA _____

 Activity 14.2 A Quick Guide to Solving Genetics Problems

Over the years, rules have been developed for setting up genetics problems and denoting
genes and their alleles in these problems. This activity provides a quick review of some
of these rules. After you have read through all of this material, complete Activities 14.3,
14.4, and 14.5.

Basic Assumptions to Make When Solving Genetics Problems

1. Are the genes linked?
If the problem does not (a) indicate that the genes are linked or (b) ask whether the genes
are (or could be) linked, then you should assume that the genes are not linked.

2. Are the genes sex-linked?
Similarly, if the problem does not (a) indicate that the genes are sex-linked (that is, on
the X chromosome) or (b) ask whether the genes are (or could be) on the X chromosome
(or Y chromosome), then you should assume that the genes are on autosomes and are not
sex-linked.

3. Is there a lethal allele?
If a gene is lethal, then you should assume that the offspring that get the lethal allele
(if dominant) or alleles (if homozygous recessive) do not appear; that is, they are not
born, do not hatch, and so on. Therefore, they are not counted among the offspring.
(An obvious exception is lethal genes that have their effect late in life. If this is the case,
however, it should be noted in the question.)

4. Are the alleles dominant, recessive, or neither?
Unless the problem states otherwise, assume that capital letters (*BB*, for example)
designate dominant alleles and lowercase letters (*bb*, for example) indicate recessive
alleles. When there is codominance or incomplete dominance, the alleles are usually
designated by the same capital letter and each one is given a superscript (for example,
$I^A I^B$ in Figure 14.15 on page 273 of *Biological Science*, 5th edition).

5. How are genotypes written?
Assume a gene for fur color in hamsters is located on the number 1 pair of homologous
autosomes. Brown fur (*B*) is dominant over white fur (*b*). The genotype for fur color can
be designated in different ways:

a. The alleles can be shown associated with the number 1 chromosome. In this
   notation, an individual heterozygous for this gene is designated as $|^B|^{b.}$

b. Most commonly, this notation is simplified to *Bb*.

In problems that involve sex-linked genes, the chromosomes are always indicated—for example, $X^A X^a$ and $X^a Y$.

6. **What information do you need to gather before trying to solve a genetics problem?**

Before trying to solve any problem, answer these questions:

a. What information is provided? For example, what type of cross is it? Is it a monohybrid or dihybrid cross? Are the genes sex-linked or autosomal? Linked or unlinked?

b. What does the information provided tell you about the gene(s) in question? For example:
   • What phenotypes can result?
   • How many alleles does the gene have?
   • Are the alleles of the gene dominant? Recessive? Codominant?

c. Does the question supply any information about the individuals' genotypes? If so, what information is provided?
   • Grandparent information?
   • Parental (P) information?
   • Gamete possibilities?
   • Offspring possibilities?

Solving Genetics Problems

1. What is a Punnett square?

Punnett squares are frequently used in solving genetics problems. A Punnett square is a device that allows you to determine all the possible paired combinations of two sets of characteristics. For example, if you wanted to determine all the possible combinations of red, blue, and green shirts with red, blue, and green pants, you could set up this Punnett square:

		Shirts		
		Red shirt	Blue shirt	Green shirt
Pants	Red pants			
	Blue pants			
	Green pants			

Similarly, if you wanted to determine the probability of a male (XY) and a female (XX) having a son or a daughter, you would first determine the possible gametes each could produce and then set up a Punnett square to look at all the possible combinations of male and female gametes. Here meiosis dictates that the female gametes get one of her X chromosomes or the other. In the male, the gametes get either the X chromosome or the Y. As a result, the Punnett square would look like this:

		Female's gamete possibilities	
		X	X
Male's gamete possibilities	X	XX	XX
	Y	XY	XY

2. If you know the parents' genotypes, how can you determine what types of offspring they will produce?

   a. **Autosomal genes:** For an autosomal gene that has the alleles *A* and *a*, there are three possible genotypes: *AA*, *Aa*, and *aa*.

   All possible combinations of matings and offspring for two individuals carrying the autosomal gene with alleles *A* and *a* are shown in the figure below.

   If you know how to solve these six crosses, you can solve any problem involving one or more autosomal genes.

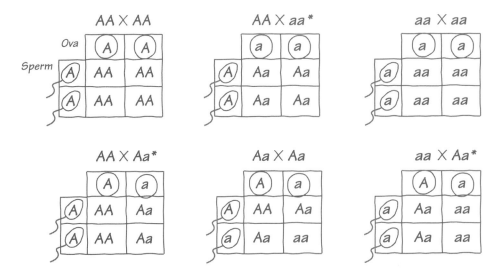

Note: If you take sex into account, there are actually nine possible combinations of matings:

Male genotypes	Female genotypes		
	AA	*Aa*	*aa*
AA	*AA* × *AA*	*AA* × *Aa*	*AA* × *aa*
Aa	*Aa* × *AA*	*Aa* × *Aa*	*Aa* × *aa*
aa	*AA* × *aa*	*aa* × *Aa*	*aa* × *aa*

Because the results of reciprocal autosomal matings (e.g., *AA* male with *aa* female and *aa* male with *AA* female) are the same, those matings are highlighted in the nine combinations above.

b. **Sex-linked genes:** For sex-linked genes, females have three possible genotypes: $X^{w+}X^{w+}$, $X^{w+}X^w$, and X^wX^w. Males have only two possible genotypes: $X^{w+}Y$ and X^wY. All the possible combinations of matings and offspring for a sex-linked trait are listed in the next figure. If you know how to solve these six single-gene crosses, then you can solve any genetics problem involving sex-linked genes.

All possible combinations of matings and offspring for two individuals with a sex-linked gene are shown in the figure below.

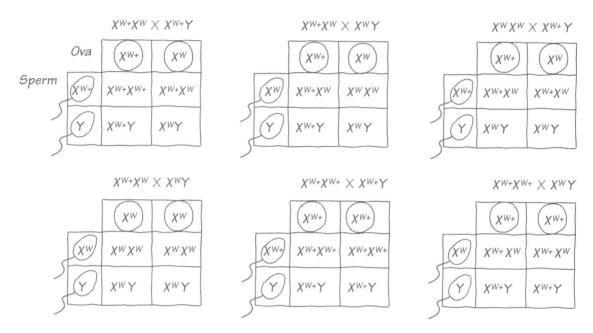

c. **Multiple genes:** Remember, if genes are on separate chromosomes, then they assort independently in meiosis. Therefore, to solve a genetics problem involving multiple genes, where each gene is on a separate pair of homologous chromosomes:
- Solve for each gene separately.
- Determine probabilities for combination (multiple-gene) genotypes by multiplying the probabilities of the individual genotypes.

Example:

What is the probability that two individuals of the genotype *AaBb* and *aaBb* will have any *aabb* offspring?

To answer this, solve for each gene separately.
A cross of *Aa* × *aa* could produce the following offspring:

	A	a
A	a	aa
A	Aa	aa

½ *Aa* and ½ *aa* offspring

A cross of *Bb* × *Bb* could produce the following offspring:

	B	b
B	BB	Bb
B	Bb	bb

¼ *BB*, ½ *Bb* and ¼ *bb* offspring

The probability of having any *aabb* offspring is then the probability of having any *aa* offspring multiplied by the probability of having any *bb* offspring.

The probability is ½ × ¼ = ⅛.

Activity 14.3 How can you determine all the possible types of gametes?

To solve genetics problems in which genotypes are given, you must first know what types of gametes each organism can produce.

1. How many different kinds of gametes can individuals with each of the following genotypes produce?

   a. *AA*

   b. *aa*

   c. *Aa*

   d. *AaBB*

   e. *AaBb*

   f. *AaBbCC*

   g. *AaBbCc*

   h. *AaBbCcDdEeFf*

2. Based on your answer in question 1, propose a general rule for determining the number of different gametes organisms like those described in question 1 can produce.

3. Two individuals have the genotypes *AaBbCcDd*.

   a. How many different types of gametes can each produce?

   b. What are these gametes?

   c. You set up a Punnett square using all the possible gametes for both individuals. How many "offspring squares" are in this Punnett square?

   d. If you completed this Punnett square, how easy would it be to find all the "offspring squares" that contain the genotype *AaBBccDd*?

   e. Given that the genes are all on separate pairs of homologous chromosomes, what other method(s) could you use to determine the probability of these individuals having any offspring with the genotype *AaBbccDd*?

Activity 14.4 Solving Problems When the Genetics Are Known

Refer to Activity 14.3 and to Chapter 14 in *Biological Science*, 5th edition, to complete this activity.

1. An organism that has the genotype *AaBbCc* is crossed with an organism that has the genotype *AABbCc*. Assume all genes are on separate sets of chromosomes (that is, they are not linked).

   a. What is the probability that any of the offspring will have the genotype *AABBCC*? (*Hint:* To get the answer, consider the six possible types of autosomal crosses. Determine the individual probabilities of getting *AA* offspring from the monohybrid cross. Then do the same to determine the probabilities of getting *BB* offspring and *CC* offspring. Multiply these probabilities together.)

   b. What is the probability that any of the offspring will have the genotype *AaBbcc*?

2. Consider the cross *AaBbCcddEe* × *AABBccDDEe*.

   a. What is the probability that any offspring will have the genotype *AaBBCcDdEE*?

   b. What is the probability that any offspring will have the genotype *AABBCCDDee*?

3. In cattle, primarily Holsteins, there is an X-linked disease that causes hairlessness with tooth problems; the disease is called anhidrotic ectodermal dysplasia (AED). Affected cattle have little hair, few sweat glands, and no incisors.

   a. What is the probability that a heterozygous female Holstein mated with an affected male will produce any affected (AED) offspring?
   $X^{a+} X^a \times X^a Y$ = heterozygous female crossed with an AED male (a^+ = unaffected; a = affected)

b. What is the probability that the mating in part a will produce any affected females?

c. What is the probability that this mating will produce any diseased males?

4. A heterozygous brown-eyed human female who is a carrier of color blindness marries a blue-eyed male who is not color-blind. Color blindness is a sex-linked trait. Assume that eye color is an autosomal trait and that brown is dominant over blue. What is the probability that any of the offspring produced have the traits listed?

a. brown eyes

b. blue eyes

c. color blindness

d. color-blind males

e. brown-eyed, color-blind males

f. blue-eyed, color-blind females

g. What is the probability that any of the males will be color-blind?

h. Why do males show sex-linked traits more often than females?

Name _____   Course/Section _____

Date _____   Professor/TA _____

Activity 14.5 Solving Problems When the Genetics Are Unknown

An understanding of Mendelian genetics allows us to determine the theoretical probabilities associated with normal transmission of autosomal and sex-linked alleles during reproduction. This understanding provides us with strategies for solving genetics problems. In real-life situations, geneticists use these strategies to determine the genetics behind specific phenotypic traits in organisms. They do this by conducting controlled crosses of experimental organisms (e.g., Drosophila) or by analyzing family pedigrees (e.g., humans).

Part One: Controlled Crosses

Two problems are presented below. In each, you are given:

(a) "Wild population"—the phenotypic characteristics of a wild population of fruit flies that were trapped randomly on a remote island.

(b) "Cross 1, 2, etc."—the phenotypic characteristics of offspring from a controlled cross. The phenotypes of the parents are indicated after each cross (e.g., "Cross 1: Male Ambler × Female Wild Type").

For each of the problems, analyze the results in each cross and answer the questions that follow.

Problem 1

Wild population	Wild type	Ambler	Total
Male	33	17	50
Female	31	19	50
Total	64	36	100

Cross 1: Male Ambler × Female Wild Type

Offspring Vial 1	Wild type	Ambler	Total
Male	29	24	53
Female	29	31	50
Total	58	55	113

a. What does cross 1 tell you about dominance versus recessiveness of the alleles?

b. What does cross 1 tell you about placement of the alleles on autosomes versus sex chromosomes? (If you can determine this, your answer must show the chromosomal genotypes for the parents in this cross.)

Cross 2: Female Ambler × Male Wild Type

Offspring Vial 2	Wild type	Ambler	Total
Male	0	32	32
Female	32	0	32
Total	32	32	64

a. What does cross 2 tell you about dominance versus recessiveness of the alleles?

b. What does cross 2 tell you about placement of the alleles on autosomes versus sex chromosomes? (In your answer, show the chromosomal genotypes for the parents in this cross.)

Problem 2
Mt = Monocle; Bt = Bifocal; Tr = Trifocal; Sp = Spinner; Sh = Shiny

Wild Population	Mt, Sp	Mt, Sh	Bt, Sp	Bt, Sh	Tr, Sp	Tr, Sh	Total
Male	10	6	6	0	22	3	46
Female	19	1	9	1	20	4	54
Total	29	6	15	1	42	7	100

Cross 1: Bifocal, Spinner Female × Monocle, Shiny Male

Mt = Monocle; *Bt* = Bifocal; *TR* = Trifocal; *Sp* = Spinner; *Sh* = Shiny

Offspring Vial 1	*Mt, Sp*	*Mt, Sh*	*Bt, Sp*	*Bt, Sh*	*Tr, Sp*	*Tr, Sh*	Total
Male	0	0	0	0	31	34	65
Female	0	0	0	0	34	38	72
Total	0	0	0	0	65	72	137

a. What does cross 1 tell you about dominance versus recessiveness of the alleles?

b. What does cross 1 tell you about placement of the alleles on autosomes versus sex chromosomes?

Cross 2: Monocle, Spinner Female × Trifocal, Spinner Male

Mt = Monocle; *Bt* = Bifocal; *Tr* = Trifocal; *Sp* = Spinner; *Sh* = Shiny

Offspring Vial 2	*Mt, Sp*	*Mt, Sh*	*Bt, Sp*	*Bt, Sh*	*Tr, Sp*	*Tr, Sh*	Total
Male	8	8	0	0	8	8	32
Female	23	0	0	0	15	0	38
Total	31	8	0	0	23	8	70

a. What does cross 2 tell you about dominance versus recessiveness of the alleles?

b. What does cross 2 tell you about placement of the alleles on autosomes versus sex chromosomes?

Part Two: Analysis of Pedigrees

Analyze the pedigree and answer the questions that follow.

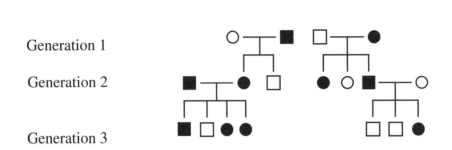

The diagram above shows a pedigree of three generations in a family. Filled circles and squares indicate persons with a genetic disorder. A square indicates a male, and a circle indicates a female. The two males in generation 1 are siblings, in this case, brothers.

1. Looking only at the generation 2 offspring (of the two generation 1 brothers), what can you say about the gene(s) controlling the genetic disorder? For example, is the disorder caused by a gene that is dominant, recessive, autosomal, or sex-linked?

2. What additional information do you gain from examining the generation 3 offspring?

Name _____   Course/Section _____

Date _____   Professor/TA _____

 Activity 14.6 How can the mode of inheritance be determined experimentally?

Outline the experimental crosses you would need to make to solve each problem.

1. Three new traits have been discovered in a population of *Drosophila:*
    *   Tapping (a behavioral mutant in which the fly taps one foot constantly)
    *   Single stripe (a pigmentation change that leads to a long stripe down the fly's back)
    *   Angular (causes angular bends in bristles that are normally straight)

    The positions of the three genes on the chromosomes are unknown. Given two pure-breeding (homozygous) lines and using an initial cross of normal, normal, normal females with tapping, single stripe, angular males, describe the appropriate genetic experiments needed to establish whether any of these traits are caused by genes that are:

    a. Autosomal or sex-linked

    b. Linked on the same chromosome or unlinked

2. A genetics student chose a special project involving a three-gene cross to check the relative positions and map distances separating three genes in *Drosophila* that she thought were all on the third chromosome. To do this, she mated *Drosophila* females that were homozygous for the recessive genes *cu* (curled), *sr* (striped), and *e* (ebony) with males that were homozygous for the wild type, cu^+ (straight), sr^+ (not striped), and e^+ (gray). She then mated (testcrossed) the F_1 females with homozygous recessive curled, striped, ebony males.

Here are the phenotypic results of the testcross:

straight, gray, not striped	820
curled, ebony, striped	810
straight, ebony, striped	100
curled, gray, not striped	97
straight, ebony, not striped	80
curled, gray, striped	90
straight, gray, striped	1
curled, ebony, not striped	2
Total:	2000

a. How are the three genes arranged on the chromosomes?

b. What evidence allows you to answer the question in part a?

c. If any of the genes are linked, how far apart are they on the chromosome? How can you determine this?

Name _____   Course/Section _____

Date _____   Professor/TA _____

 Activity 15.1 Is the hereditary material DNA or protein?

Accumulating and Analyzing the Evidence

Build a concept map to review the evidence used to determine that DNA was the genetic material, the structure of DNA, and its mode of replication. Recall from Activity 4.1 and Chapter 4 in *Biological Science*, 5th edition, how Watson and Crick determined the structure of DNA. Keep in mind that there are many ways to construct a concept map. Use the following guidelines:

- First, develop a separate concept map for each set of terms (A to D on the next page). Begin by writing each term on a separate Post-it note or sheet of paper.
- Then organize each set of terms into a map that indicates how the terms are associated or related.
- Draw lines between the terms, and add action phrases to the lines to indicate how the terms are related.

Here is an example:

- After you have completed each of the individual concept maps, merge or interrelate the maps to show the overall logic used to conclude that DNA (not protein) is the heredity material.
- When you have completed the overall concept map, answer the questions.

TERMS:

Map A

Griffith

mice

S strain of *Streptococcus*

R strain of *Streptococcus*

live

heat-killed

transformation

Avery, McCarty, and MacLeod

DNA

protein

Map B

Hershey and Chase

bacteria

bacteriophage (phage) (only a protein
and DNA)

^{35}S

^{32}P

Waring blender

high-velocity centrifugation

Map C

Watson and Crick

X-ray crystallography

Chargaff's rule

purine structure

pyrimidine structure

H bonds

phosphate sugar backbone

Map D

Meselson and Stahl

conservative

dispersive

semiconservative

nucleic acid bases

^{14}N

^{15}N

bacteria

density equilibrium centrifugation

replication

1. In the early to mid-1900s, there was considerable debate about whether protein or DNA was the hereditary material.

   a. For what reasons did many researchers assume that protein was the genetic material?

b. What key sets of experiments led to the understanding that, in fact, DNA and not protein was the hereditary material?	c. What evidence did each experiment provide?
Griffith experimented with a lethal S strain of bacteria and a nonlethal R strain of the same bacterial species. S strain injected into mice killed them. R strain didn't, and neither did heat-killed S strain. If heat-killed S and live R strains were mixed and injected simultaneously, however, the mice died. When autopsied, live S bacteria were found.	

2. Watson and Crick were the first to correctly describe the structure of DNA. What evidence did they use to do this? How did they use this evidence to put together or propose the structure of DNA?

3. How did the results of Meselson and Stahl's experiments show that DNA replicates semiconservatively? To respond to this, answer the following questions.

   a. Diagram the results that would be expected for each type of replication proposed.

   b. What evidence allowed Meselson and Stahl to eliminate the conservative model?

   c. What evidence allowed them to eliminate the dispersive model?

Name _____   Course/Section _____

Date _____   Professor/TA _____

Extend Your
Understanding   **Activity 15.2 How do bacteria replicate their DNA?**

An *E. coli* cell that contains a single circular chromosome made of double-stranded DNA is allowed to replicate for many generations in an ^{15}N medium until all of the *E. coli* cells' DNA is labeled with ^{15}N (nitrogen 15). One *E. coli* cell is removed from the ^{15}N medium and is placed into medium in which all of the available nucleotides are ^{14}N labeled. The *E. coli* cell is allowed to replicate until eight *E. coli* are formed.

1. Given this scenario, which of the following statements is true?

    a. Some ^{15}N-labeled DNA will be found in all eight cells.

    b. Some ^{15}N-labeled DNA will be found in only four of the cells.

    c. Some ^{15}N-labeled DNA will be found in only two of the cells.

    d. Some ^{15}N-labeled DNA will be found in only one of the cells.

2. Draw the sequence of events that occurred to explain your answer.

Name _____   Course/Section _____

Date _____   Professor/TA _____

 Activity 15.3 How does DNA replicate?

Working in groups of three or four, construct a dynamic (working or active) model of DNA replication. You may use the materials provided in class or devise your own.

Building the Model

- Develop a model of a short segment of double-stranded DNA.
- Include a key for your model that indicates what each component represents in the DNA molecule—for example, adenine, phosphate group, deoxyribose.
- Create a dynamic (claymation-type) model of replication. Actively move the required bases, enzymes, and other components needed to model replication of your DNA segment.

Your model should describe the roles and relationships of all the following enzymes and structures in replication:

parental DNA	$3' \rightarrow 5'$ versus $5' \rightarrow 3'$
nucleotide excision repair	nitrogenous bases A, T, G, C
daughter DNA	replication fork
mutation	replication bubble
antiparallel strands	Okazaki fragment
single-stranded DNA-binding proteins	DNA polymerase
leading strand	helicase
telomeres	DNA ligase
lagging strand	primase
telomerase	RNA primers
5′ end	origin of replication
3′ end	

Use your model to answer these questions.

1. How did Meselson and Stahl's experiments support the idea that DNA replication is semiconservative?

2. A new form of DNA is discovered that appears to be able to replicate itself both in the 3′ → 5′ direction and in the 5′ → 3′ Effect on amino acid sequence direction. If this is true, how would this newly discovered DNA replication differ from DNA replication as we know it?

3. Amazingly, an alien species of cellular organism is found alive in the remains of a meteorite that landed in the Mojave Desert. As a scientist, you are trying to determine whether this alien life-form uses DNA, protein, or some other type of compound as its hereditary material.

   a. What kinds of experiments would you propose to determine what the hereditary material is?

   b. If the hereditary material turns out to be similar to our DNA, describe the simplest experiments you could run to try to determine if it is double-stranded like our DNA, triple-stranded, or some other configuration.

4. Some researchers estimate that the mutation rate for any given gene (or its DNA) in certain strains of bacteria is about 10^{-8}. This means that one error or mutation in a given gene is introduced for every 100 million cell divisions.

   a. What can cause mistakes in replication?

   b. How are such mistakes normally corrected?

Name _____   Course/Section _____

Date _____   Professor/TA _____

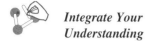 *Integrate Your Understanding* **Chapters 16–17—Modeling Transcription and Translation: What processes produce RNA from DNA and protein from mRNA?**

Create a model of the processes of transcription and translation. Your model should be a dynamic (working or active) representation of the events that occur first in transcription in the nucleus and then in translation in the cytoplasm.

When developing and explaining your model, be sure to include definitions or descriptions of the following terms and structures:

gene	start codon (methionine)
DNA	aminoacyl-tRNA synthetase
nucleotides: A, T, G, and C versus A, U, G, and C	amino acids (see Figure 16.6 on page 311 in *Biological Science*, 5th edition)
RNA modification(s) after transcription	peptidyl transferase
mRNA	polypeptide
RNA polymerase	energy
poly(A) tail	codons
5′ effect on amino acid sequence cap	stop codons
	anticodons
translation	initiation
protein synthesis	elongation
ribosome (large versus small subunit)	termination
A, P, and E sites	polypeptide
tRNA	
rRNA	

For the purposes of this activity, assume there are *no introns* in the mRNA transcript.

Building the Model

- Use chalk on a tabletop, or a marker on a large sheet of paper, to draw a cell's plasma membrane and nuclear membrane. The nucleus should have a diameter of about 12 inches.
- Draw a DNA molecule in the nucleus that contains the following DNA sequence:

  Template strand   3'—TAC TTT AAA GCG ATT—5'

  Nontemplate strand   5'—ATG AAA TTT CGC TAA—3'
- Use playdough or cut-out pieces of paper to represent the various enzymes, ribosome subunits, amino acids, and other components.
- Use the pieces you assembled to build a dynamic (claymation-type) model of the processes of transcription and translation.
- When you feel you have developed a good working model, use it to explain the processes of transcription and translation to another student or to your instructor.

Use your model of transcription and translation to answer the following questions.

1. How would you need to modify your model to include intron removal? Your explanation should contain definitions or descriptions of the following terms and structures:

   pre-mRNA   exons

   RNA splicing   spliceosome

   introns

2. If 20% of the DNA in a guinea pig cell is adenine, what percentage is cytosine? Explain your answer.

Integrate Your Understanding Chapters 16–17

3. Several different types of RNA exist in prokaryotic and eukaryotic cells. List the three main types of RNA involved in transcription and translation. Answer the questions to complete the chart.

a. Types of RNA:	b. Where are they produced?	c. Where and how do they function in cells?

4. Given your understanding of transcription and translation, fill in the blanks below and indicate the 5′ and 3′ ends of each nucleotide sequence. Again, assume no RNA processing occurs.

Nontemplate strand of DNA: 5′ A T G T A T G C C A A T G C A 3′

Template strand of DNA: _′ T _ _ _ _ _ _ _ _ _ _ _ _ _′

mRNA   :   _′ A _ _ _ _ U _ _ _ _ _ _ _ _′

Anticodons on complementary tRNA: _′ _ _ _ / _ _ _ / _ _ _ / _ _ _ / _ _ _ / _′

5. Scientists struggled to understand how four bases could code for 20 different amino acids. If one base coded for one amino acid, the cell could produce only four different kinds of amino acids (4^1). If two bases coded for each amino acid, there would be four possible choices (of nucleotides) for the first base and four possible choices for the second base. This would produce 4^2 or 16 possible amino acids.

a. What is the maximum number of *three-letter codons* that can be produced using only four different nucleotide bases in DNA?

b. How many different codons could be produced if the codons were four bases long?

Mathematical logic indicates that at least three bases must code for each amino acid. This led scientists to ask:

- How can we determine whether this is true?
- Which combinations of bases code for each of the amino acids?

To answer these questions, scientists manufactured different artificial mRNA strands. When placed in appropriate conditions, the strands could be used to produce polypeptides.

Assume a scientist makes three artificial mRNA strands:

    (x) 5′—AAAAAAAAAAAAAAAAAAAAAAAAAAA—3′
    (y) 5′—AAACCCAAACCCAAACCCAAACCCAAA—3′
    (z) 5′—AUAUAUAUAUAUAUAUAUAUAUAUAU—3′

When he analyzes the polypeptides produced, he finds that:

    x produces a polypeptide composed entirely of lysine.
    y produces a polypeptide that is 50% lysine and 50% proline.
    z produces a polypeptide that is 50% isoleucine and 50% tyrosine.

c. Do these results support the three-bases-per-codon or the four-bases-per-codon hypothesis? Explain.

d. This type of experiment was used to discover the mRNA nucleotide codons for each of the 20 amino acids. If you were doing these experiments, what sequences would you try next? Explain your logic.

6. Now that the complete genetic code has been determined, you can use the strand of DNA shown here and the codon chart in Figure 15.6 (page 284) in your textbook to answer the next questions.

Original template strand of DNA: 3′—TAC GCA AGC AAT ACC GAC GAA—5′

a. If this DNA strand produces an mRNA, what is the sequence of the mRNA?

b. For what sequence of amino acids does this mRNA code? (Assume it does not contain introns.)

c. The chart lists five point mutations that may occur in the original strand of DNA. What happens to the amino acid sequence or protein produced as a result of each mutation? (*Note:* Position 1 refers to the first base at the 3′ end of the transcribed strand. The last base in the DNA strand, at the 5′ end, is at position 21.)

Original template strand: 3′—TAC GCA AGC AAT ACC GAC GAA—5′

Mutation	Effect on amino acid sequence
i. Substitution of T for G at position 8	
ii. Addition of T between positions 8 and 9	
iii. Deletion of C at position 15	
iv. Substitution of T for C at position 18	
v. Deletion of C at position 18	
vi. Which of the mutations produces the greatest change in the amino acid sequence of the polypeptide coded for by this 21-base-pair gene?	

7. Sickle-cell disease is caused by a single base substitution in the gene for the beta subunit of hemoglobin. This base substitution changes one of the amino acids in the hemoglobin molecule from glutamate to valine. Look up the structures of glutamate (glu) and valine (val) in Figure 3.2 on page 43 of your textbook. What kinds of changes in protein structure might result from this substitution? Explain.

8. Why do dentists and physicians cover patients with lead aprons when they take mouth or other X-rays?

9 to 13. During DNA replication, which of the following observations would you expect to be true? Explain your answers.

T/F 9. More ligase would be associated with the lagging strand than with the leading strand.

T/F 10. More primase would be used for the lagging strand than for the leading strand.

T/F 11. More helicase would be associated with the lagging strand than with the leading strand.

T/F 12. DNA ligase links the 3′ end of one Okazaki fragment to the 5′ Effect on amino acid sequence of the other Okazaki fragment in the lagging strand.

T/F 13. In the lagging strand, the enzyme DNA polymerase III that produces the next Okazaki fragment also removes the short segment of primer RNA on the previous Okazaki fragment.

14. You obtain a sample of double-stranded DNA and transcribe mRNA from this DNA. You then analyze the base composition of each of the two DNA strands and the one mRNA strand, and get the following results. The numbers indicate the percentage of each base in the strand:

	A — G — C — T — U
strand 1	40.1 – 28.9 – 9.9 – 0.0 – 21.1
strand 2	21.5 – 9.5 – 29.9 – 39.1 – 0.0
strand 3	40.0 – 29.0 – 9.7 – 21.3 – 0.0

a. Which of these strands must be the mRNA? Explain.

b. Which one is the template strand for the mRNA? Explain.

15. In a new experiment, you obtain a different sample of double-stranded DNA and transcribe mRNA from this DNA. You then analyze the base composition of each of the two DNA strands and the one mRNA strand, and get the following results. The numbers indicate the percentage of each base in the strand:

$$A - G - C - T - U$$

	A	G	C	T	U
strand 1	29.1	39.9	31.0	0.0	0.0
strand 2	0.0	30.9	39.8	30.2	0.0
strand 3	29.4	39.4	31.2	0.0	0.0

a. Which of these strands could be the mRNA? Explain.

b. Which one must be the template strand for the mRNA? Explain.

Name _____   Course/Section _____

Date _____   Professor/TA _____

Activity 18.1 How is gene expression controlled in bacteria?

Fill in the chart to organize what we know about the *lac* and *trp* operons.

Operon:	*lac*		*trp*	
Is the metabolic pathway anabolic or catabolic?				
What regulatory genes are associated with the operon, and what functions does each serve?	Genes:	Functions:	Genes:	Functions:

(continued)

Operon:	lac		trp	
What structural genes are included in each operon, and what does each produce?	Genes:	Products:	Genes:	Products:
Is the operon inducible or repressible?				
Is the repressor protein produced in active or inactive form?				
The repressor protein becomes active when it interacts with:				

Activity 18.2 Modeling the *lac* and *trp* Operon Systems: How can gene expression be controlled in prokaryotes?

Using the information in Activity 18.1 and in Chapter 18 of *Biological Science,* 5th edition, construct a model or diagram of the normal operation of both the *lac* and *trp* operon systems.

In your models or diagrams, be sure to include these considerations:

> regulatory and structural genes
> inducible versus repressible control
> anabolic versus catabolic enzyme activity
> negative versus positive controls

Use your model to answer these questions.

1. Under what circumstances would the *lac* operon be "on" versus "off"? The *trp* operon?

2. How are the *lac* and *trp* operons similar (in structure, function, or both)?

3. What are the key differences between the *lac* and *trp* operons?

4. What advantages are gained by having genes organized into operons?

5. Strain X of *E. coli* contains a mutated *lac* regulatory gene on its bacterial genome. As a result, the gene produces a nonfunctional *lac* repressor protein. You add a plasmid (an extra circular piece of double-stranded DNA) to these cells. The plasmid contains a normal regulatory gene and a normal *lac* operon.

Build a model or diagram of what one of these modified *E. coli* cells would look like. Then answer the questions and use your model or diagram to explain your answers.

a. Before the addition of the plasmid, would the *E. coli* strain X cells be able to produce the enzymes for lactose digestion? Explain.

b. After the addition of the plasmid, would the plasmid's *lac* operon produce the enzymes for lactose digestion constitutively (all the time) or only when lactose was the available sugar source? Explain.

c. After the addition of the plasmid, would the bacterial genome's *lac* operon produce the enzymes for lactose digestion constitutively or only when lactose was the available energy source? Explain.

d. If equal amounts of lactose and glucose were present in the cell, would the *lac* operon in the bacterial DNA be off or on? Would the *lac* operon on the introduced plasmid be off or on? Explain.

Name _____   Course/Section _____

Date _____   Professor/TA _____

Activity 19.1 How is gene activity controlled in eukaryotes?

Human genes cannot all be active at the same time. If they were, all the cells in our bodies would look the same and have the same function(s). For specialization to occur, some genes or gene products must be active while others are turned off or inactive.

1. In eukaryotes, gene expression or gene product expression can be controlled at several different levels. Indicate what types of control might occur at each level of gene or gene product expression.

Level	Types of control
a. The gene or DNA itself	
b. The mRNA product of the gene	
c. The protein product of the mRNA	

2. Single-celled organisms such as *Amoeba* and *Paramecia* often live in environments that change quickly. Which of the following types of control allow organisms like *Amoeba* to respond most quickly to frequent short-term environmental changes? Explain your reasoning.

   a. Control of mRNA transcription from DNA

   b. Control of enzyme concentration by controlling the rate of mRNA translation

   c. Control of activity of existing enzymes

   d. Control of the amount of DNA present in the cell

Name _____   Course/Section _____

Date _____   Professor/TA _____

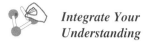 *Integrate Your Understanding* **Chapters 20–21 (1)—How and why are genes cloned into recombinant DNA vectors?**

Develop a model to explain how a human gene can be cloned into a bacterial plasmid. Your model should be a dynamic (working or active) representation of the events that need to occur in order to

- clone the insulin gene into a bacterial plasmid, and
- transform the plasmid into *E. coli.*

When you develop and explain your model, be sure to include definitions or descriptions of the following terms and components:

restriction enzyme(s)

ligase

plasmid DNA

human mRNA (the mRNA for insulin)

reverse transcriptase

transformation

E. coli marker genes (an antibiotic resistance gene)

cloning vector

Building the Model

- Use chalk on a tabletop, or a marker on a large sheet of paper, to draw at least two test tubes and an *E. coli* cell's plasma membrane. The *E. coli* should have a diameter of at least 12 inches. The test tubes should have a width of at least 6 inches. Use the test tubes for producing the insulin gene and for cloning the gene into the plasmid. Then transform your recombinant plasmid into the *E. coli* cell.
- Use modeling clay or cut-out pieces of paper to represent the enzymes, RNA molecules, and other components.

- Use the pieces you assembled to develop a dynamic (claymation-type) model to demonstrate how a gene can be cloned into a plasmid and how the plasmid can then be transformed into a bacterial cell.
- When you feel you have developed a good working model, demonstrate it to another student or to your instructor.

Use your model to answer the following questions.

1. Prior to recombinant gene technology, the insulin required to treat diabetes was obtained from the pancreases of slaughtered farm animals. Because the insulin was from other species, some humans developed immune responses or allergic reactions to it. As recombinant gene technology advanced, researchers explored the possibility of incorporating the human insulin gene into a plasmid that could be transformed into *E. coli*. If this technology was successful, the *E. coli* would produce human insulin that could be harvested from the bacterial culture medium.

   Researchers first needed to isolate the gene for insulin. To do this, they isolated mRNA (rather than DNA) from the beta cells of human pancreas tissue. Using reverse transcriptase, they made double-stranded DNA molecules that were complementary to the mRNA molecules they extracted from the pancreas cells.

   a. Based on what you know about eukaryotic chromosomes and genes, why did researchers choose to isolate mRNA rather than DNA?

   b. What further adjustments might researchers need to make in the DNA molecules produced by reverse transcriptase before the molecules could be incorporated into bacterial plasmids?

   c. Not all of the DNA molecules produced by reverse transcription from pancreatic mRNA contained the gene for insulin. Some contained other genes. What mechanisms can be used to locate those bacterial colonies that picked up plasmids containing any of the genes produced by reverse transcription from pancreatic mRNAs?

d. What mechanisms can be used to locate bacterial colonies that picked up only plasmids containing the insulin gene? (*Note:* Today, almost all of the insulin used for the treatment of human diabetes is produced using recombinant technology.)

Name _____   Course/Section _____

Date _____   Professor/TA _____

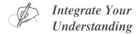 *Integrate Your Understanding* **Chapters 20–21 (2)—How can PCR be used to amplify specific genes?**

1. Assume you are using PCR (polymerase chain reaction) to make multiple copies of a gene (shaded in gray below).

   DNA containing gene of interest:

   3′–TATAAAGACTTACAAATTTGTCCCCATTTTGC–5′

   5′–ATATTTCTGAATGTTTAAACAGGGGTAAAACG–3′

On separate sheets of paper, describe the overall process and diagram the results you would obtain for 1, 2, and 3 rounds of PCR replication using the primers ATGTT and CCATT.

(*Note:* For simplicity we are showing DNA primers that are only 5 bases in length. In actual use, the DNA primers used are at least 17 bases long. This length is used to help reduce the risk that the primer anneals with—base-pairs with—anything other than the specific segment of DNA to be amplified.)

2. PCR is often used in forensics to amplify small amounts of DNA found at crime scenes. The amplified DNA is then tested for differences in RFLP (restriction fragment length polymorphisms) or STR (single tandem repeat) lengths.

   a. Explain what RFLPs and STRs are.

   b. How do STRs compare for unrelated individuals versus for closely related individuals (for example, parent and child or brother and sister)?

   c. How reliable are these types of DNA fingerprinting for identifying individuals? What factors affect their reliability?

3. Answer the following questions regarding the use of DNA technology in forensics.

    a. After undergoing electrophoresis, the gel in the figure below shows the RFLP analysis of DNA samples obtained from a crime scene. Blood stains on a suspect's shirt (B) were analyzed and compared with blood from the victim (V) and from the suspect (S). Are the blood stains on the shirt from the victim or from the suspect? Explain.

b. Investigators find an additional blood sample in the garage, but they are not sure whether it is related to the crime. After undergoing electrophoresis, the gel in the figure below shows the results of DNA fingerprinting of blood samples obtained from the victim (V), her husband (H), the suspect (S), and the blood from the garage (B_2). Is the blood from the garage related to the crime? Explain your answer.

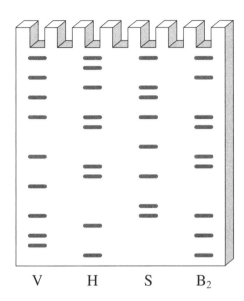

4. Which of the following sequences (on one strand of a double-stranded DNA molecule) is likely to be a cleavage site for a restriction enzyme? Explain your answer.

   a. CGTACC

   b. ATGTCG

   c. GATATG

   d. TGCGCA

   e. TGGCCG

Name _____   Course/Section _____

Date _____   Professor/TA _____

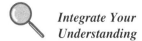 *Integrate Your Understanding* **Chapters 20–21 (3)—How can we discover the sequence of an organism's DNA?**

Bacterial genomes have between 1 and 6 million base pairs (Mb). Most plants and animals have about 100 Mb. Humans have approximately 2900 Mb. As a result, individual chromosomes may contain millions of base pairs. It is difficult to work with DNA sequences this large. As a result, the DNA is broken into smaller pieces (approximately 500 to 1000 bp each). These pieces are sequenced and then the sequenced pieces are examined and aligned based on overlapping sequence homology at their ends.

By comparing the DNA sequences among organisms, scientists can determine:

- what parts of the genomes are most similar among organisms and therefore likely evolved earliest,
- what key changes exist in the genomes that may account for differences among related species, and
- what changes within species exist that may account for development of specific types of disease.

The following activity has been designed to help you understand how genomes are sequenced and how the sequence information may be used.

1. In 1980, Frederick Sanger was awarded the Nobel Prize for inventing the dideoxy method (or Sanger method) of DNA sequencing. A double-stranded DNA segment approximately 700 bp in length is heated (or treated chemically) to separate the two strands. The single-stranded DNA that results is placed into a test tube containing a 9-to-1 ratio of normal deoxynucleotides to dideoxynucleotides. A dideoxynucleotide has no –OH group at either the 2′ or 3′ carbon. As a result, whenever a dideoxynucleotide is added to the growing DNA strand, synthesis stops at that point. If the ratio of normal to dideoxynucleotides is high enough, where one will be included in the sequence is random.

Using this method, you set up each of four test tubes as noted below:

Tube number	Deoxynucleotide	Dideoxynucleotide
1	dATP, dTTP, dGTP, dCTP	ddATP
2	dATP, dTTP, dGTP, dCTP	ddTTP
3	dATP, dTTP, dGTP, dCTP	ddGTP
4	dATP, dTTP, dGTP, dCTP	ddCTP

All tubes contain the same single-stranded DNA molecules and the same primers. All other enzymes and materials that are required for DNA replication are present in each tube. You allow the replication to continue for the same length of time in each tube. At the end of the time period, you extract the DNA from each tube and run it on an agarose gel. You dye the gel with ethidium bromide and observe the following banding patterns on the gel.

(*Note:* For this demonstration, we are using a DNA strand that is only 20 bp in length.)

a. Which band in the gel contains the shortest DNA strand? What is the identity of its terminal ddNTP?

         Integrate Your Understanding Chapters 20–21 (3)

b. Which band contains the next shortest DNA strand? What is the identity of its terminal ddNTP?

c. Continue reading the terminal ddNTP of each band from shortest to longest to determine the linear sequence of nucleotides in the DNA strand complement. What is the sequence?

2. The Sanger method has since been modified so that each ddNTP used is now flagged with an identifying fluorescent tag such that ddTTP fluoresces red; ddGTP, yellow; ddCTP, blue; and ddATP, green.

    a. Assume that you run the same experiment that you did above. However, this time you combine all of the different nucleotides (both dNTPs and ddNTPs) in the same test tube. You run the products of the reaction on the following gel. Draw the bands you would see on the gel using the appropriate fluorescent colors (ddTTP fluoresces red; ddGTP, yellow; ddCTP, blue; and ddATP, green).

b. Draw the image you would see from an optical scan of the gel, as is shown in Figure 20.9 on page 377 of *Biological Science*, 5th edition.

c. Assume that this is a gene from a normal individual. Compare the optical image you generated in part b to the optical image from an individual with a hypothetical genetic disease. This disease results from nonproduction of a necessary gene product. What and where is the mutation in the gene sequence? Why might this particular mutation lead to nonproduction of a protein?

3. To help determine evolutionary relationships among different groups of organisms, scientists compare gene sequences of highly conserved genes. What are "highly conserved genes"? Give examples and indicate what is "highly conserved" and why.

4. What types of DNA do scientists use to determine individual identities of organisms within the same species? Why do they use this type of DNA?

Name _____   Course/Section _____

Date _____   Professor/TA _____

 Activity 22.1 How can the development of an organism be controlled at the cellular and molecular levels?

The critical events that occur in the development of all multicellular organisms include zygote formation, cell division, induction, determination and differentiation, and morphogenesis.

1. Describe the function(s) of each of the following events in development.

Stage of development	General function
a. Zygote formation	
b. Cell division	
c. Differential gene expression	
d. Determination and differentiation	
e. Morphogenesis	

2. *Drosophila melanogaster*, the fruit fly, has been used as a model organism to study development. As a result of these studies (and others), we now realize that normal development involves hierarchies of spatially and temporally regulated gene expression.

    a. What general approach(es) did researchers take to find the genes responsible for normal development in *Drosophila*? For example, what types of observations did they make, and what method(s) did they use?

3. Using these methods, researchers have discovered that development occurs due to the sequential expression of a variety of genes. Many of these are genes for transcription factors. As Nüsslein-Volhard and Wieschaus discovered, in *Drosophila*, the basic sequence or flow of development can be outlined as shown below.

       Maternal effects genes (for example, *bicoid*)
          \
         Gap genes
             \
            Pair-rule genes
                \
               Segment polarity genes
                   \
                  Homeotic genes (or homeobox, *Hox*, genes)
                     \
                    Master regulatory genes (for example, *myoD*)

    a. What does each type of gene control in the development of the fruit fly?

b. Based on your answers in part a, if you wanted to produce a mutation that caused profound effects (drastic effects) on the overall development of the organism, where in the sequence would you introduce the mutation? Explain your answer.

c. What phenotype change(s) might you see if a developing fruit fly had a mutation in a gap gene? In a segment polarity gene?

4. What, if any, similarities and differences are found when *Drosophila* development is compared with development in other organisms, such as a frog, a mouse, or a human? What do the findings indicate about the evolution of these organisms?

Extend Your
Understanding   **Activity 22.2 An Experiment Designed to Test the Genetics**
                    of Development

To look for new genes affecting development, a researcher treats male fruit flies with a mutagen. In the P generation, each of the males is then mated individually with a normal untreated female. F_1 offspring from each cross are mated with each other, again with only one male and one female per mating vial. F_2 offspring from each of these crosses are again mated with each other as above. The resulting F_3 offspring are again crossed as above.

What results would you expect from these matings if a mutant development phenotype were the result of:

    a. a dominant mutant allele

    b. a recessive mutant allele

    c. mutant maternal genetic input to the egg

Name _____   Course/Section _____

Date _____   Professor/TA _____

Activity 23.1 What major events occur in animal development?

1. In biology, males and females are defined based on the types of gametes they produce. Males produce smaller, motile gametes; females produce larger, nonmotile gametes.

    In your text, sperm are described as "DNA with propellers" or "race cars." In contrast, eggs are described as "semitrailers." What aspects of the structure and function of (a) sperm and (b) eggs make these good analogies?

    a. Sperm:

    b. Eggs:

2. The early stages in the development of all vertebrates (and many other animal species) include zygote formation, cleavage, blastula formation, gastrula formation, and organogenesis.

    Among the major aspects of development are cell adhesion/recognition, cell growth (in size and/or number), cell determination, and cell differentiation.

    Fill in the chart on the next page to indicate:

    a. What key events occur at each of the stages of development?

    b. Which of the major aspects of development are involved in each stage of development?

Stages of development	Zygote formation	Cleavage and blastula formation	Gastrula formation	Organogenesis
a. Summary of key events for each stage				
b. Aspects of development that occur during this stage:				
Cell adhesion/ recognition				
Cell growth (size or number?)				
Determination				
Differentiation				

Activity 23.1

Name _____   Course/Section _____

Date _____   Professor/TA _____

*Extend Your
Understanding*   **Activity 23.2 Animal Development: Sample Exam Questions**

Questions 1 and 2. For each situation described, choose the graph that fits best.

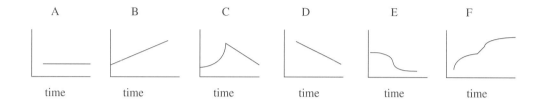

1. Which graph best describes the change in size of individual cells of a vertebrate embryo from the time of zygote formation to the end of cleavage? Explain your answer.

2. Which graph best describes the change in the number of cells in the embryo from fertilization to gastrulation? Explain your answer.

Questions 3 to 6. Although the mechanics of gastrulation may differ among various organisms, the overall objectives and problems are the same. Which of the following are accomplished by the end of gastrulation?

T/F   3. formation of the three germ layers (i.e., ectoderm, mesoderm, and endoderm)

T/F   4. establishment of the embryonic axis (anterior to posterior)

T/F   5. determination of the fates of the individual cells of the gastrula

T/F   6. formation of the neural tube

7. In a now classic experiment, Spemann and Mangold took cells from the dorsal lip of the blastopore from one frog embryo and transplanted them to an area opposite the dorsal lip in another frog embryo (the host). As a result, the host embryo developed two heads, one at the site of the "new" or transplanted dorsal lip and the other at the site of the host's original dorsal lip. A thorough examination indicated that *the second head was formed from host cells*. Which of the following developmental cues or mechanisms was most likely the trigger (or cause) for the generation of the second head? Explain your answer.

A. cytoplasmic determinants, which are unequally distributed in the host embryo
B. gradients set up in the egg, which gave rise to the host embryo
C. *Hox* genes (or *homeobox* genes) present in the embryo receiving the transplant
D. cell–cell signaling between the transplanted dorsal lip and the host embryo
E. all of the above

Name _____   Course/Section _____

Date _____   Professor/TA _____

 Activity 24.1 What major events occur in plant development?

1. What key differences exist in form and function of plants versus animals?

2. The fundamental differences that exist in plant cell structure also dictate that plant development must differ from animal development. Which aspects of plant development differ significantly from those of animal development and which are similar?

Note: To give yourself enough space for the answers, re-create the table below on your computer or on a larger sheet of paper.

Aspect of plant development	Differences from animal development	Similarities to animal development
Gamete formation		
Zygote formation		
Embryogenesis		
Vegetative development		
Reproductive development		

Name _____   Course/Section _____

Date _____   Professor/TA _____

Extend Your
Understanding **Activity 24.2 The Developmental Genetics of Floral Structure Mutations**

Questions 1 and 2. In studying the development of flowers, Meyerowitz and colleagues found three general classes of floral mutants:

- Mutants with only carpels and stamens
- Mutants with only sepals and carpels
- Mutants with only petals and sepals

From this they developed their ABC model of how three genes interacted under normal conditions to form the four whorls of the flower: sepals, petals, stamens, and carpel.

In their model, each of the three organ identity genes is expressed in two adjacent whorls such that if the *A* gene's protein is produced alone, sepals are formed. If the proteins from both the *A* and *B* genes are produced, petals are formed. If the proteins from genes *B* and *C* are both produced, stamens are formed. If only the protein from gene *C* is produced, carpels are formed (see Figure 24.10 on page 440 of *Biological Science,* 5th edition).

Given the mutants (listed above) that they found, they also proposed that the presence of A protein inhibits C protein production and the presence of C protein similarly inhibits the production of A protein.

1. What floral phenotype would have resulted if the *A* gene was lacking and no inhibition between A and C protein production existed?

2. What floral phenotype would have resulted if the *C* gene was lacking and no inhibition between A and C protein production existed?

3. What floral phenotype would have resulted if the *B* gene was lacking and no inhibition between A and C protein production existed?

4. Compare your answers to 1–3 above to the drawings of the actual mutants in Figure 23.12 on page 410 of *Biological Science,* 4th edition.

    a. How do they differ?

    b. Does proposing the mutual inhibition of *A* and *C* gene production explain these actual mutants?

Name _____   Course/Section _____

Date _____   Professor/TA _____

Activity 25.1 How did Darwin view evolution via natural selection?

Darwin is remembered not because he was the first to propose that evolution occurs. Many others had presented this idea before. Instead, he is remembered for defining the mechanism behind evolution—that is, the theory of natural selection. To do this, Darwin integrated, or put together, information from a wide range of sources. Some of this information was provided by others; some he observed on his own.

Working alone or in groups of three or four, construct a concept map of Darwin's view of evolution via natural selection. Be sure to include definitions or descriptions of all the terms in the list below. Keep in mind that there are many ways to construct a concept map, and consider the following guidelines:

- Begin by writing each term on a separate Post-it note or piece of paper.
- Then organize the terms into a map that indicates how the terms are associated or related.
- Draw lines between terms, and add action phrases to the lines that indicate how the terms are related.
- When you finish your map, explain it to another group of students.

Here is an example:

$$\boxed{Embryology} \xrightarrow{\text{provides evidence for}} \boxed{Evolution}$$

Terms

fact	Darwin	fit individuals
biogeography	vertebrate limb structure	fossil record
gradualism	species population	embryology
uniformitarianism	individual	taxonomy
theory	variability	selective (domestic)
Galápagos Islands	paleontology	breeding
evolution	Malthus	limited resources
homology	population size	struggle for existence
natural selection	environment or	reproduction
analogy	resources	extinction

Activity 25.1   165

Use the understanding you gained from creating the concept map to answer the following questions.

1. In the 1860s, what types of evidence were available to indicate that evolution had occurred on Earth?

2. How did knowledge of mechanisms of artificial selection (used in developing various strains of domesticated animals and plants) help Darwin understand how evolution could occur?

3. Based on his studies, Darwin made some observations; they are listed in the following chart. Complete the chart by answering how Darwin made the observations.

Observation	What did Darwin read or observe that led him to this observation?
a. All species populations have the reproduction potential to increase exponentially over time.	
b. The number of individuals in natural populations tends to remain stable over time.	
c. Environmental resources are limited.	
d. Individuals in a population vary in their characteristics.	
e. Much of this variation is heritable.	

4. Based on these observations, Darwin made some inferences. Which of the observation(s) in question 3 allowed Darwin to make each inference?

Inference	Observations that led to the inference
a. Production of more individuals than the environment can support leads to a struggle for existence such that only a fraction of the offspring survive each generation.	
b. Survival for existence is not random. Those individuals whose inherited traits best fit them to the environment are likely to leave more offspring than less fit individuals do.	
c. The unequal ability of individuals to survive and reproduce leads to a gradual change in the population, with favorable characteristics accumulating over the generations.	

5. Based on these observations and inferences, how did Darwin define fitness?

6. How did Darwin define evolution?

7. What is the unit of natural selection? (What is selected?) What is the unit of evolution? (What evolves?)

8. In a population of mice, some individuals have brown fur and some have black fur. At present, both phenotypes are equally fit. What could happen to change the relative fitness of the two phenotypes in the population? For example, what could cause individuals with brown fur to show reduced fitness relative to individuals with black fur?

9. Assume you discover a new world on another planet that is full of organisms.

   a. What characteristics would you look for to determine that these organisms arose as a result of evolutionary processes?

   b. What characteristics would you look for to determine that these organisms did *not* arise as the result of evolutionary processes?

10. Why is it incorrect to say that vertebrates evolved eyes in order to see?

Name _____   Course/Section _____

Date _____   Professor/TA _____

 Activity 25.2 How do Darwin's and Lamarck's ideas about evolution differ?

Early in the 1800s Lamarck proposed a theory of evolution. He suggested that traits acquired during an organism's life—for example, larger muscles—could be passed on to its offspring. The idea of inheritance of acquired characteristics was popular for many years. No such mechanism is implied in Darwin's theory of evolution via natural selection, however. After Darwin published his work, scientists conducted many experiments to disprove the inheritance of acquired traits. By the middle of the 20th century, enough data had accumulated to make even its most adamant supporters give up the idea of inheritance of acquired characteristics.

Given your understanding of both Lamarck's and Darwin's ideas about evolution, determine whether the statements on the next page are more Lamarckian or more Darwinian. If the statement is Lamarckian, change it to make it Darwinian. Here are two example statements and answers.

Examples

A. The widespread use of DDT in the mid-1900s put pressure on insect populations to evolve resistance to DDT. As a result, large populations of insects today are resistant to DDT.

Answer: This is a Lamarckian statement. DDT worked only against insects that had no DDT-resistance genes. The genes for DDT resistance had to be present for insects to survive DDT use in the first place.

Suggested change: Wide-scale use of DDT in the mid-1900s selected against insects that had no resistance to DDT. Only the insects that were resistant to DDT survived. These insects mated and passed their resistance genes on to their offspring. As a result, large populations of insects today are resistant to DDT.

B. According to one theory, the dinosaurs became extinct because they couldn't evolve fast enough to deal with climatic changes that affected their food and water supplies.

Answer: This is a quasi-Lamarckian statement. Organisms do not purposefully evolve. (Genetic recombination experiments are perhaps an exception.) Once you are conceived, your genes are not going to change; that is, you are not going to evolve. The genetic composition of a species population can change over time as certain genotypes are selected against. Genes determine phenotypes. The environmental conditions may favor the phenotype produced by one genotype more than that produced by another.

Suggested change: According to one theory, the dinosaurs became extinct because their physiological and behavioral characteristics were too specialized to allow them to survive the rapid changes in climate that occurred. The climatic changes caused changes in the dinosaurs' food and water supplies. Because none of the dinosaurs survived, the genes and associated phenotypes that would have led to their survival must not have been present in the populations.

Statements

1. Many of the bacterial strains that infect humans today are resistant to a wide range of antibiotics. These resistant strains were not so numerous or common prior to the use of antibiotics. These strains must have appeared or evolved in response to the use of the antibiotics.
2. Life arose in the aquatic environment and later invaded land. Once animals came onto land, they had to evolve effective methods of support against gravity and locomotion in order to survive.

3. A given phenotypic trait—for example, height, speed, tooth structure—(and therefore the genes that determine it) may have positive survival or selective value, negative survival or selective value, or neutral (neither positive nor negative) survival or selective value. Which of these it has depends on the environmental conditions the organism encounters.

4. The children of bodybuilders tend to be much more athletic, on average, than other children because the characteristics and abilities gained by their parents have been passed on to the children.

Name _____   Course/Section_____

Date _____   Professor/TA_____

 Activity 25.3 How would you evaluate these explanations of Darwin's ideas?

Unfortunately, even today some people get or give the impression that acquired characteristics can be inherited. As a result, we need to be very careful about how we state our understanding of evolution and evolutionary theory.

To test understanding of Darwin's ideas, the following question was included on an exam.

4-point question:
In two or three sentences, describe Darwin's theory of descent with modification and the mechanism, natural selection, that he proposed to explain how this comes about.

Four student answers to the question are given. Based on what you know about Darwinian evolution and natural selection, evaluate and grade how well each answer represents Darwin's ideas. For any answer that does not receive full credit (4 points), be sure to indicate why points were lost.

Student 1. Darwin saw that populations increased faster than the ability of the land to support them could increase, so that individuals must struggle for limited resources. He proposed that individuals with some inborn advantage over others would have a better chance of surviving and reproducing offspring and so be naturally selected. As time passes, these advantageous characteristics accumulate and change the species into a new species.

Student 2. Darwin's theory of evolution explains how new species arise from already existing ones. In his mechanism of natural selection, organisms with favorable traits tend to survive and reproduce more successfully, while those that lack the traits do not. Beneficial traits are passed on to future generations in this manner, and a new species will be created in the end!

Student 3. Descent with modification using natural selection was Darwin's attempt at explaining evolution. An organism is modified by its surroundings, activities, and lifestyle. These modifications, by natural selection, make the organism better suited to its life.

Student 4. Darwin's theory states that organisms can become modified by environmental conditions or use or disuse features and that the modifications can be passed down to succeeding generations. He proposes that nature selects for a characteristic trait that is beneficial to the survival of the organisms and that organisms would pass on this trait.

Activity 25.3

Name _____   Course/Section _____

Date _____   Professor/TA _____

Activity 26.1 A Quick Review of Hardy–Weinberg Population Genetics

Part A. Review Chapter 26 of *Biological Science*, 5th edition. Then complete the discussion by filling in the missing information.

If evolution can be defined as a change in gene (or more appropriately, allele) frequencies, is it conversely true that a population not undergoing evolution should maintain a stable gene frequency from generation to generation? This was the question that Hardy and Weinberg answered independently.

1. **Definitions.** Complete these definitions or ideas that are central to understanding the Hardy–Weinberg theorem.

    a. Population: An interbreeding group of individuals of the same _____.
    b. Gene pool: All the alleles contained in the gametes of all the individuals in the
        _____.

    c. Genetic drift: Evolution (defined as a change in allele frequencies) that occurs in
        _____ populations as a result of chance events.

2. **The Hardy–Weinberg theorem.** The Hardy–Weinberg theorem states that in a population that (**is/is not**) evolving, the allele frequencies and genotype frequencies remain constant from one generation to another.

3. **Assumptions.** The assumptions required for the theorem to be true are listed in Chapter 26 on page 467 of your textbook and are presented here in shortened form.

    a. There are no random _____ that would affect the gene in question.
    b. There is no net _____ of individuals into or out of the population.
    c. There is no net _____; that is, the forward and backward mutation rates for alleles are the same. For example, *A* goes to *a* as often as *a* goes to *A*.
    d. Mating is at _____ for the trait/gene(s) in question.
    e. There is no _____. Offspring from all possible matings for the trait/gene are equally likely to survive.

4. **The Hardy–Weinberg proof.** Consider a gene that has only two alleles, R (dominant) and r (recessive). The sum total of all R plus all r alleles equals all the alleles at this gene locus, or 100% of all the alleles for that gene.

Let p = the percentage or probability of all the R alleles in the population.
Let q = the percentage or probability of all the r alleles in the population.

If all R + all r alleles = 100% of all the alleles, then

$$p + q = 1 \text{ (or } p = 1 - q \text{ or } q = 1 - p)$$

Note: Frequencies are stated as percentages (for example, 50%) and their associated probabilities are stated as decimal fractions (for example, 0.5).

Assume that 50% of the alleles for fur color in a population of mice are B (black) and 50% are b (brown). The fur-color gene is autosomal.

a. What percentage of the gametes in the females (alone) carries the B allele?

b. What percentage of the gametes in the females (alone) carries the b allele?

c. What percentage of the gametes in the males carries the B allele?

d. What percentage of the gametes in the males carries the b allele?

e. Given the preceding case and all the Hardy–Weinberg assumptions, calculate the probabilities of the three possible genotypes (RR, Rr, and rr) occurring in all possible combinations of eggs and sperm for the population.

Female gametes and probabilities		
	$(R)(p)$	$(r)(q)$
Male gametes and probabilities $(R)(p)$	_RR_ (p^2)	_Rr_ (pq)
$(r)(q)$	_Rr_ (pq)	_rr_ (q^2)

Because the offspring types represent all possible genotypes for this gene, it follows that $p^2 + 2pq + q^2 = 1$ or 100% of all genotypes for this gene.

Part B. Use your understanding of the Hardy–Weinberg proof and theorem to answer these questions.

1. According to the Hardy–Weinberg theorem, $p + q = 1$ and $p^2 + 2pq + q^2 = 1$. What does each of these formulas mean, and how are the formulas derived?

2. Assume a population is in Hardy–Weinberg equilibrium for a given genetic autosomal trait. What proportion of individuals in the population are heterozygous for the gene if the frequency of the recessive allele is 1%?

3. About one child in 2500 is born with phenylketonuria (an inability to metabolize the amino acid phenylalanine). This is known to be a recessive autosomal trait.

   a. If the population is in Hardy–Weinberg equilibrium for this trait, what is the frequency of the phenylketonuria allele?

   b. What proportion of the population constitutes carriers of the phenylketonuria allele (i.e., what proportion is heterozygous)?

4. In purebred Holstein cattle, about one calf in 100 is spotted red rather than black. The trait is autosomal, and red is recessive to black.

   a. What is the frequency of the red alleles in the population?

   b. What is the frequency of black homozygous cattle in the population?

   c. What is the frequency of black heterozygous cattle in the population?

5. Assume that the probability of a sex-linked gene for color blindness is $q = 0.09$ and the probability of the normal allele is $p = 0.91$. This means that the probability of X chromosomes carrying the color-blindness allele is 0.09 and the probability of X chromosomes carrying the normal allele is 0.91.

   a. What is the probability of having a color-blind male in the population?

   b. What is the probability of a color-blind female?

6. The ear-tuft allele in chickens is autosomal and produces feathered skin projections near the ear on each side of the head. This gene is dominant and is lethal in the homozygous state. In other words, homozygous dominant embryos do not hatch from the egg. Assume that in a population of 6000 chickens, 2000 have no ear tufts and 4000 have ear tufts. What are the frequencies of the normal versus ear-tuft alleles in this population?

7. How can one determine whether or not a population is in Hardy–Weinberg equilibrium? What factors need to be considered?

8. Is it possible for a population's genotype frequencies to change from one generation to the next but for its gene (allele) frequencies to remain constant? Explain by providing an example.

Name _____   Course/Section _____

Date _____   Professor/TA _____

Activity 26.2 Violating the Assumptions of the Hardy–Weinberg Theorem

In each of the following scenarios, determine which assumption of the Hardy–Weinberg theorem is being violated. Explain your answer.

1. In a particular region of the coast, limpets (a type of mollusk) live on nearshore habitats that are uniformly made up of brown sandstone rock. The principal predators of these limpets are shorebirds. The limpets occur in two morphs, one with a light-colored shell and one with a dark-colored shell. The shorebirds hunt by sight and are able to see the light ones on the dark sandstone easier than the dark ones.

2. In *Chen caerulescens* (a species of goose), the white body form—the snow goose—and the blue body form—the blue goose—occasionally coexist. In these areas of contact, white-by-white and blue-by-blue matings are much more common than white-by-blue matings.

3. Prior to the Mongolian invasions that occurred between the 6th and 16th centuries, the frequency of blood type B across Europe was close to zero. The frequency of blood type B among the Mongols was relatively high. Today, it is possible to see fairly high frequencies of blood type B in the eastern European countries and a gradual decrease in the frequency of blood type B as one moves from the eastern European countries to the western European countries (e.g., France and England).

Name _____   Course/Section _____

Date _____   Professor/TA _____

Activity 26.3 What effects can selection have on populations?

1. What effects can natural selection have on populations? For example, what types of selection can occur in a population, and how are these defined?

2. Examine the scenarios on the following pages. For each scenario:

   a. Decide whether or not natural selection is operating. In doing this, indicate whether there is variability in the population(s).

      If *no*, what does this imply about evolution?
      If *yes*, what is the nature of the variation? For example, what characteristics must the variation have for selection to operate on it?

   b. Is there any indication that members of the population(s) differ in fitness?
      If *no,* what does this imply about the operation of natural selection?
      If *yes,* describe the difference in fitness.

   c. Given your answers to parts a and b, what trend should characterize the future behavior or composition of the population(s)?

   Note: Be sure to indicate any assumptions you make in answering the questions.

Scenario I. A particular species of mouse feeds on the seeds of a single species of cherry tree. When the mice eat a seed, they digest it completely. The mice choose seeds of intermediate and large sizes, leaving the very small seeds of the cherry tree uneaten.

   a.

   b.

   c.

   d. Graph the abundance of the different sizes of cherry tree seeds over 10 seasons. Explain your rationale for drawing your graph this way.

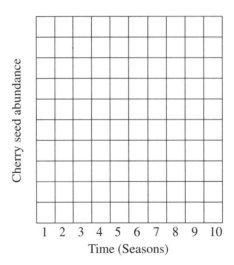

Scenario IIa. Small island A contains three separate populations of a single species of cherry tree. The seed size varies between trees. That is, some trees produce seeds that are all in the small size ranges, others produce seeds all in the middle size ranges, and others produce seeds in the large size ranges. A small population of mice is introduced to the island. The mice eat cherries and are the only predators of the cherry trees. When the mice eat a cherry, they completely digest it and the pit or seed inside it. The mice choose medium and large seeds and leave the smallest seeds uneaten.

a.

b.

c.

d. Graph the abundance of the different sizes of cherry tree seeds over 10 seasons on island A. Explain your rationale for drawing your graph this way, and indicate how this might impact the population sizes of the different types of cherry trees on the island over time.

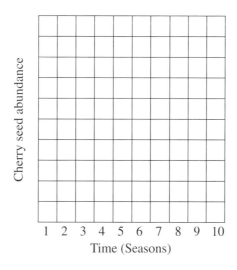

Scenario IIb. Would your answer for Scenario IIa change given the following information? Explain.

As you continue to study the populations of trees, you note that the viability of the seeds varies with size such that the viability of the small seeds is less than that of the middle-sized seeds, which is less than that of the largest seeds.

a.

b.

c.

Scenario IIIa. Small island B contains three separate populations of a single species of cherry tree. Unlike the species on island A, in this species seed size varies within trees. That is, each tree produces seeds that range in size from large to small. A small population of mice is introduced to the island. These are the only predators of the cherry trees. When the mice eat a cherry, they digest it and the pit or seed completely. The mice choose medium and large seeds and leave the smallest seeds uneaten.

a.

b.

c.

d. Graph the abundance of the different sizes of cherry tree seeds over 10 seasons on island B. Explain your rationale for drawing your graph this way, and indicate how this might impact the population sizes of the different types of cherry trees on the island over time.

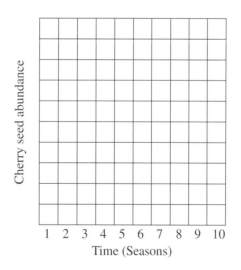

Scenario IIIb. Would your answer for Scenario IIIa change given the following information? Explain.

As you continue to study these populations, you note that the viability of the seeds does not vary with size. Over time, however, you find that the trees that grow from the smallest seeds produce fewer large seeds.

a.

b.

c.

Scenario IV. After a severe spring ice storm, about half of the finches (small birds) in a population are found dead. Examination of the dead birds indicates that they vary in age from young to old. About 60% of the dead are fledglings (just left the nest); about 20% are over 3 years of age (old for this species).

a.

b.

c.

3. The figure below shows the distribution of a phenotypic characteristic in a population of organisms. Also shown are three portions (A, B, and C) of that distribution.

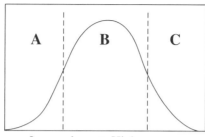

Low or less → High or more

Use the following set of answers:
A. Portion A only
B. Portion B only
C. Portion C only
D. Portions A and C simultaneously
E. Portions B and C simultaneously

Parts a, b, and c. For the following situations, which portion(s) of the population will be selected *against* and be *least likely* to have its/their genes represented in the next generation? Explain your answers.

a. Male sticklebacks with bright red coloring are favored by female sticklebacks as mates. However, the bright red color makes the males more likely to be seen by predators.

b. On island Z in the Galápagos, the plant population contains only two species. One of the two plant species produces very large seeds and the other produces small seeds. A species of seed-eating finch has lived on the island for many years. Members of this established species have large beaks and prefer large seeds. A small population of a different species of seed-eating finch has migrated to the island. The beak size in the new species' population varies over a continuum from relatively small to large. How will evolution of the new finch species be affected?

c. In a species of plant-eating land snail, the shell color is variable and ranges from very light to intermediate to very dark in color. The snails are preyed upon by birds that use sight to find their prey. A small population of these snails is moved to an island where the food plants in their preferred habitat are either very light in color or very dark.

Extend Your
Understanding **Activity 26.4 Evolution of Non-Calling Crickets on Kauai**

Dr. Marlene Zuk and colleagues have observed an interesting case of evolution in action on the Hawaiian island of Kauai in populations of male crickets that have stopped calling, in order to avoid a parasitoid fly, *Ormia ochracea.* The flies follow the chirps of a calling cricket and then lay fly larvae, maggots, onto the cricket's back. The maggots burrow into the cricket and emerge a week later—killing the cricket in the process. The crickets of Kauai have evolved a strategy to avoid parasitism by the flies: they don't call and can therefore avoid detection by the fly. At first, researchers thought this was a behavioral adaptation in the crickets, but when they looked more closely, they found that the non-calling males lacked the noisemaking wing structures that normal males have, and are therefore unable to call. Researchers observed this evolutionary change in less than 20 generations, which is remarkably fast to see such a large change. This adaptation, although it keeps the crickets safe from parasitism, poses another problem. Male crickets call in order to find a female mate. If the males can't call, how can they mate? What will happen to the cricket population on Kauai if the crickets can't mate?

Researchers found that approximately 90% of the male crickets on Kauai lost the ability to call, but 10% of the male cricket population retained the ability to call. These calling males began to attract not only females, but non-calling males as well, which would often intercept a female before she was able to reach the calling male. Females would normally not choose to mate with a non-calling male; but on Kauai, researchers think that females have evolved somewhat relaxed preferences in response to the prevalence of non-calling males in the population.

1. In what situations would you expect the silent-wing mutation to be favored by natural selection? In what situations would you expect the normal-wing version to be favored by natural selection? List at least two ideas for each.

2. What do you think will happen to the population of crickets on Kauai? State your prediction and your null hypothesis.

3. How would you design an experiment to test your hypothesis?

4. Draw a graph of the data that would support your hypothesis.

Activity 26.4

Name _____   Course/Section _____

Date _____   Professor/TA _____

 Activity 27.1 What factors affect speciation?

1. The Galápagos Archipelago consists of a dozen islands, all within 64 km of their
    nearest neighbor. From 1 to 11 of the 13 species of Darwin's finches live on each
    island. Many evolutionary biologists believe that if there had been only one island,
    there would be only one species of finch. This view is supported by the fact that
    Cocos Island is isolated (by several hundred kilometers of open ocean) from the
    other islands in the archipelago, and only one species of finch is found there.

    a. How does the existence of an archipelago promote speciation? Explain or
        provide an example.

    b. Is the mode of speciation that occurred on these islands more likely to have been
        allopatric or sympatric? Explain.

    c. Is the type of speciation seen on the Galápagos Archipelago more likely to be the
        result of anagenesis or cladogenesis? Explain.

2. Hybrids formed by mating two different species are often incapable of reproducing successfully with each other or with the members of their parent populations. Explain why this is the case. (*Hint:* Consider what you know about chromosome numbers and meiosis.)

3. Because most hybrids can't reproduce, their genes (and the genes of their parents) are removed from the population. Only the genes of individuals that breed with members of their own species remain in the population. This implies that there is a strong selective advantage for genes that enable individual organisms to recognize members of their own species. Today, a wide range of reproductive isolating mechanisms has been identified. Each of the following scenarios describes a reproductive isolating mechanism. Indicate whether each is a prezygotic or postzygotic isolating mechanism. Explain your answers.

   a. Crickets use species-specific chirp patterns to identify a mate of their own species.

   b. Two species of butterfly mate where their ranges overlap and produce fertile offspring, but the hybrids are less viable than the parental forms.

c. Two species of a plant cannot interbreed, because their flowers differ in size and shape and require pollination by different species of bee.

d. Two species of firefly occupy the same prairie and have similar flash patterns, but one is active for a half-hour around sunset, whereas the other doesn't become active until an hour after sunset.

4. Many of our most successful grain crops arose as hybrids; most are also allopolyploids. These crops can successfully reproduce. Explain.

Name _____   Course/Section _____

Date _____   Professor/TA _____

Activity 27.2 How does hybridization affect speciation?

Two species of toads, *Bombina bombinna* and *Bombina variegata*, share a hybrid zone that is 4000 km in length and only 10 km wide. The frequency of alleles specific to *B. bombinna* decreases from close to 100% on one edge of the hybrid zone to 0% on the opposite edge. Similarly, the frequency of *B. variegata*–specific alleles decreases (beginning at the opposite edge of the hybrid zone) from close to 100% to 0% across the hybrid zone.

1. On the graph below, map the approximate percentage of each type of species-specific gene across the hybrid zone. Use an X to indicate the frequency of *B. bombinna*–specific genes and an O to indicate the frequency of *B. variegate*–specific genes.

2. How would you predict this gene distribution would change over time if:

a. reinforcement occurs?

b. fusion occurs?

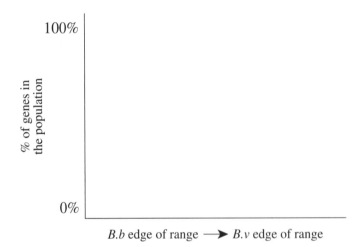

c. Hypothetically, which of the above events (a or b) is more likely to occur if all environmental conditions across the two species ranges are similar? Explain.

d. Hypothetically, which event (a or b) is more likely to occur if the environmental conditions vary gradually across the species ranges such that one end of the range was much colder than the other, for example? Explain.

Name _____ Course/Section _____

Date _____ Professor/TA _____

Integrate Your
Understanding **Units 3 and 5—Why sex?**

Read the following statement. Then, based on your knowledge of genetics and evolution, decide to agree or disagree with the statement. Whichever you decide, write a short paragraph that provides solid evidence defending your choice.

When you think about sexual reproduction, it makes no sense. After all, evolution selects for organisms that are best fit. In a population of sexually reproducing organisms, a mutant that reproduced asexually would increase its representation more quickly than the wild type (i.e., it would have a higher fitness). So we should expect asexual reproduction to be much more widespread among eukaryotic species than it is.

Name _____   Course/Section _____

Date _____   Professor/TA _____

 Activity 28.1 How are phylogenies constructed?

Construct a modified concept map to relate the ideas of phylogeny and systematics (see Terms, below) to the cladogram on the next page. Use the following guidelines:

- Begin by writing each term on a separate Post-it note or piece of paper.
- Then indicate how the terms are associated or related to each other and to the cladogram on the next page.
- Be sure to include definitions or descriptions of all the terms as you use them to explain these relationships.

Terms

clade	monophyletic	shared derived character
cladistics	polyphyletic	out-group
cladogram	paraphyletic	in-group
homology	convergent evolution	taxonomy
analogy	shared primitive character	phylogeny

Use the understanding you gained from creating the concept map to answer the following questions.

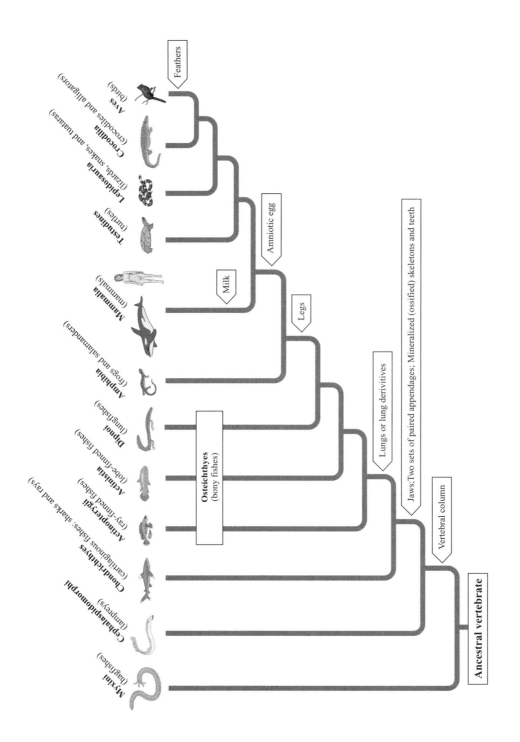

Myxini (hagfishes)

Cephalaspidomorphi (lampreys)

Chondrichthyes (cartilaginous fishes: sharks and rays)

Actinopterygii (ray-finned fishes)

Actinistia (lobe-finned fishes)

Dipnoi (lungfishes)

Amphibia (frogs and salamanders)

Mammalia (mammals)

Testudines (turtles)

Lepidosauria (lizards, snakes, and tuataras)

Crocodilia (crocodiles and alligators)

Aves (birds)

Feathers

Amniotic egg

Milk

Legs

Lungs or lung derivitives

Jaws; Two sets of paired appendages; Mineralized (ossified) skeletons and teeth

Vertebral column

Osteichthyes (bony fishes)

Ancestral vertebrate

1. Compare the taxonomy of a group with its phylogeny (in general terms).

	Taxonomy	Phylogeny
a. Definition or purpose of a taxonomy/phylogeny:		
b. Types of characters used to develop a taxonomy/phylogeny:		
c. What similarities could there be between the taxonomy of a given group and its phylogeny?		
d. What are the key differences between the taxonomy of a given group and its phylogeny?		

2. On the cladogram shown on the previous page, circle at least one monophyletic grouping. Justify your answer.

3. Considering only the individual representative organisms in the cladogram (for example, cardinal, whale, salamander), which of them can be used as good examples of analogy or convergent evolution? As good examples of homology? Explain your reasoning.

4. Which pairs of structures are analogous versus homologous?

   a. the dorsal fins of a dolphin and a salmon
   b. the flippers of a dolphin and arms of a human
   c. a rhesus monkey's tail and a human's coccyx
   d. the bracts (red leaves) of a poinsettia and the green leaves of a rose
   e. the bracts (red leaves) of a poinsettia and the red petals of a rose
   f. the scales on a reptile and the feathers on a bird
   g. a human female's ovaries and a male human's testes

5. In recent years, DNA sequence analysis has been used in developing phylogenetic relationships among organisms.

   a. What type of DNA has been used most commonly in this analysis? Why was this type chosen over others?

   b. The phylogenies developed using DNA sequence analysis may differ from those constructed using morphology and physiology. How do scientists know which phylogeny is more correct?

6. Answer parts a–c about the phylogenetic tree shown below.

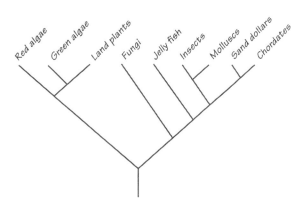

a. Which of the following is more closely related to green algae: red algae or land plants? Explain.

b. Are land plants more closely related to chordates or mollusks? Explain.

c. True or False? Jellyfish are more closely related to chordates than to fungi.

7. Answer parts a and b about the phylogenetic trees shown below.

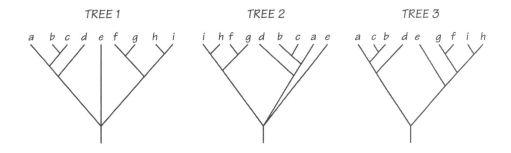

a. Identify the tree that shows a different set of relationships between the nine taxa than the other two trees. Circle the difference above.

b. If species a–h are a group of flowering plants, assume taxa a–d have red flowers, taxa f–i have blue flowers, and taxa e has white flowers. Draw the locations of flower color changes on trees 1–3. How are the three trees similar? How are they different?

Name _____   Course/Section _____

Date _____   Professor/TA _____

Activity 28.2 What is parsimony analysis?

To determine relatedness among species, parsimony analysis is often used. In Figure 28.3 of *Biological Science*, 5th edition, parsimony analysis is applied to matched DNA sequences from five different species. The same type of analysis can be applied to the morphological and developmental characteristics of species. Review Figure 28.3 and the parsimony method used. Then conduct your own parsimony analysis using the morphological characteristics of the four species of sunflowers in the following table.

1. Begin by recording the morphological characteristics of the sunflower species in the table below.

2. On the three possible phylogenies for these species (below the table), indicate the number of changes that must have occurred for each of the proposed phylogenies to be correct.

Species	Traits				
	Height (Tall/Short)	Flowers (Single/Multiple)	Leaves (Heart or Lance-shaped)	Petals (Striped or Solid)	Stem (Smooth or Hairy)
I	S	S	L	Striped	H
II	T	S	L	Solid	S
III	S	S	H	Solid	S
IV	T	M	H	Solid	H
Ancestral	S	M	H	Solid	H

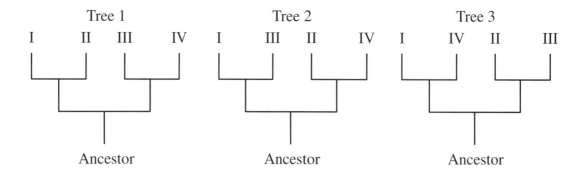

Tree 1
I   II   III   IV

Ancestor

Tree 2
I   III   II   IV

Ancestor

Tree 3
I   IV   II   III

Ancestor

3. Based on your analysis, which of the phylogenies is most parsimonious?

4. If, instead of morphology data, you used DNA data, as shown in the table below, which tree would be most parsimonious?

Species	Sites in DNA sequence				
	1	2	3	4	5
I	A	T	G	T	C
II	A	T	G	A	C
III	G	C	G	T	G
IV	C	A	C	A	C
Ancestral	A	A	G	T	C

5. Which of the proposed phylogenies (based on DNA or morphology) is more correct? Explain your answer.

Name _____   Course/Section _____

Date _____   Professor/TA _____

 ## Activity 28.3 Put Yourself in the Professor's Shoes: What questions would you ask?

One of the best ways to study for an exam is to put yourself in the professor's shoes. For example, ask yourself: What questions would I ask about the material if I were the professor? Asking and answering such questions are good practice for taking the actual exam. They also help you to better understand and organize the major ideas and concepts you have studied.

Write three exam questions designed to test how well a student understands the major concepts in Chapters 25–28. Indicate the correct answer to each question, and explain why each alternative answer is incorrect. Your questions should be of the types described next.

I. Problem solving or application of a concept or principle to a problem

For example: The little-known hypothetical organism *Skyscra parius* is a long-necked animal that feeds on the leaves of Australian trees that grow to heights of 30 feet. Being a hooved animal, *S. parius* cannot climb trees, so it feeds much like modern-day giraffes do. Fossil evidence indicates that the ancestors of *S. parius* had fairly short necks. Read the arguments presented below. For each, indicate whether or not the factors described could have affected neck length or tree height over the course of evolution of *S. parius*. (A = factors described could have affected neck length; B = factors described probably did not affect neck length.)

____ 1. *S. parius* ancestors likely demonstrated significant variation in neck length, with some having shorter necks and others having longer necks.
____ 2. When first born, juveniles of *S. parius* were much shorter than adults, so they were not able to compete successfully with adults that had longer necks.
____ 3. Female *S. parius* preferentially mate with longer-necked males.

II. Translation: The ability to recognize concepts restated in a different form or to restate concepts in a different form

For example: In what ways are the structure and function of the angiosperm seed and the amniotic egg (in this example, the chicken's egg) similar? In what ways are they different?

T/F 1. Both the angiosperm seed and the chicken's egg contain stored food for the early development of the embryo.

T/F 2. When released from the plant or laid by the chicken, both the angiosperm seed and the chicken's egg contain a partially developed diploid embryo/offspring generation.

T/F 3. The seed coat of the seed and the shell of the egg help prevent desiccation (water loss).

T/F 4. Seeds may remain dormant and viable for hundreds of years; the same is true of chick eggs.

Write your exam questions in the spaces provided.

Exam Question 1:

Answer:

Exam Question 2:

Answer:

Exam Question 3:

Answer:

Activity 28.3

Name _____   Course/Section _____

Date _____   Professor/TA _____

Integrate Your
Understanding   **Chapters 3 and 28—What do we know about the**
                  origin of life on Earth?

Construct a concept map of conditions on the early Earth and the origin of life-forms. Be sure to include definitions or descriptions of all the terms in the list below. Keep in mind that there are many ways to construct a concept map. Use the following guidelines:

- Begin by writing each term on a separate Post-it note or piece of paper.
- Then organize the terms into a map that indicates how the terms are associated or related.
- Draw lines between terms, and add action phrases to the lines that indicate how the terms are related.
- If you are doing this activity in small groups in class, explain your map to another group of students when you are done.

Here is an example:

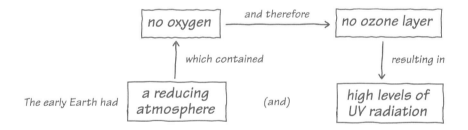

Terms

no oxygen

reducing atmosphere

high-oxygen atmosphere

sunlight

electrical discharge
(for example,
lightning)

amino acids

ozone layer

sugars

nucleic acids

DNA

inorganic compounds

organic compounds

carbon dioxide

micelle

phospholipid bilayer

ammonia

phospholipids

water

Stanley Miller

methane

molecular clock

heat

heterotrophs

autotrophs

prokaryotes

RNA world

Eukarya

Burgess Shale

Cambrian explosion

Doushantuo microfossils

Hox genes

adaptive radiation

Permian extinction

fossil record

Precambrian era

Paleozoic era

Mesozoic era

Cenozoic era

Use the understanding you gained from creating the concept map to answer the following questions.

1. Modern theory suggests that the early (pre-life) atmosphere on Earth was a reducing one. Why (for what reasons) is it believed that oxygen was not present when life formed on Earth?

2. What proposed energy sources existed on this early (pre-life) Earth?

3. In the 1950s, Stanley Miller performed a set of experiments to determine whether life could have evolved given the conditions stated in the answers to questions 1 and 2.

    a. How was the experiment designed?

    b. What were the necessary controls?

    c. What was produced in the experiment?

    d. What did the results imply about the possible origin of life on Earth?

e. There is general agreement that life must have evolved in the oceans originally and only much later invaded land. What factors of the physical environment on the early and evolving Earth support these ideas? Changes in which of these were essential to allow life to survive on land?

f. Most of us can't imagine a world without oxygen. However, as you learned earlier, chemically, oxygen is a powerful oxidizing compound. What effect(s) would the increase in oxygen levels of the atmosphere have on the organisms that existed at that time?

4. Your text discusses the fossil findings of both the Burgess Shale and the Ediacaran Hills. Why are these two deposits so interesting to biologists, geologists, and paleontologists? Explain.

5. Why is the fossil record considered to be incomplete?

Name _____   Course/Section _____

Date _____   Professor/TA _____

Chapters 9 and 28—How can we determine the age of
fossils and rocks?

To determine the age of fossils and rocks, scientists determine the amounts of radioactive
compounds and their stable daughter products present in the sample. Radioactive
elements are known to decay into stable daughter compounds at specific rates. Six
radioactive compounds, their stable daughter compounds and their half-lives, are shown
in the following table.

Radioactive compound	Stable daughter compound	Half-life
Carbon 14	Nitrogen 14	5370 years
Potassium 40	Argon 40	1.25 billion years
Rubidium 87	Strontium 87	48.8 billion years
Thorium 232	Lead 208	14 billion years
Uranium 235	Lead 207	704 million years
Uranium 238	Lead 206	4.47 billion years

For dating rocks, the potassium-argon method is often used for the following reasons:

- Argon is a gas. When rock is molten, any existing argon gas will escape. As a
  result, newly formed rocks contain no argon.
- As the ^{40}potassium naturally occurring in the rock decays to ^{40}argon, the ^{40}argon
  is trapped in pocket in the rock. The ratio of ^{40}potassium to ^{40}argon can be
  measured to determine the rock's age.

For dating organic material, carbon-nitrogen dating is often used for the following
reasons:

- The ratio of radioactive to nonradioactive carbon dioxide in the atmosphere is
  fairly constant over time. As a result, the levels of these elements in organic
  tissue remain relatively constant as long as the organism is alive.
- Once the organism is dead, no new inputs of carbon occur, and the existing ratio
  of radioactive carbon to nonradioactive carbon will decrease over time as the
  radioactive carbon decays.

1. In one half-life, half of the original radioactive compound will decay into its stable daughter compound.

    a. If a newly formed rock contains 100 units of potassium 40, how many units of potassium 40 would it contain after 1.25 billion years?

    b. How many units of argon 40 would the rock contain when newly formed versus after 1.25 billion years?

    c. After the 1.25 billion years, what would be the ratio of potassium 40 to argon 40 in the rock?

    d. After 2.5 billion years, what would be the ratio of ^{40}K to ^{40}Ar?

2. You are fossil hunting and find a trilobite fossil in an old riverbed. You have it radiometrically dated and are told it is 275 million years old.

    a. In this amount of time, how many half-lives of ^{40}K have elapsed?

    b. Given your answer in part a, what would be the ratio of ^{40}K to ^{40}Ar found in the fossil remains?

3. You want to date some fabric that you have discovered at an archeological dig.

   a. What method of dating would be best for this? Describe the general procedure for dating the cloth.

   b. Assume the cloth is 2000 years old. How would the level of radioisotope change over this period of time?

Name _____   Course/Section _____

Date _____   Professor/TA _____

Activity 29.1 What are the main characteristics of bacteria and archaea?

1. Based on DNA sequence analysis, three major domains of life have been proposed. What are the three major domains of life? What sets of characteristics place organisms into one domain versus another?

Major domains of life	Key characteristics

2. The Bacteria are divided into six major groups. What characteristics are used to place organisms into each of these groups?

Bacterial group	Characteristics
Firmicutes	
Spirochaeles	
Actinobacteria	
Chlamydiales	
Cyanobacteria	
Proteobacteria	

3. The Archaea are divided into two major groups: the Euryarchaeota and the Crenarchaeota. What characteristics are used to place organisms into each of these groups?

Archaean group	Characteristics
Euryarchaeota	
Crenarchaeota	

4. There is great diversity in the ways different species of microbes
   - obtain energy for metabolic functions, and
   - obtain carbon for building the macromolecules of life.

Fill in the chart to indicate how these characteristics are used to describe the nutritional type or nutritional classification of organisms. (Refer also to Table 29.3 on page 538 in *Biological Science*, 5th edition.)

	Prokaryotes					
Energy source						
Carbon source						
Mode of nutrition						

5. Bacteria first appear in the fossil record about 3.5 billion years ago. Humans appeared in the fossil record only a few million years ago. Given this information, which group would you say is more highly evolved?

   a. What kinds of arguments or evidence would you use to support the idea that bacteria are more highly evolved?

   b. How would you support the idea that humans are more highly evolved?

6. Given what you know about the origin of life on Earth, you want to look for life on other planets.

    a. What characteristics of a planet's environment would you look for? Explain your reasoning.

    b. What kind(s) of life would you look for? Explain your reasoning.

    c. What tests or probes would you use to find the kind of life you proposed in part a? Explain your reasoning.

Name _____   Course/Section _____

Date _____   Professor/TA _____

 Activity 29.2 How has small size affected prokaryotic diversity?

It is often said that bacteria tend to be more diverse biochemically and eukaryotes tend to be more diverse morphologically.

Answer questions 1–7. Then write a summary argument (question 8) to support the statement above. In other words, write an argument describing how small size has limited morphological diversity but promoted biochemical (metabolic) diversity among the prokaryotes.

1. All living organisms must maintain a relatively constant internal environment. Maintaining this environment means that a certain concentration of each substance must be maintained per unit volume of the cell. The cell's ability to maintain a specific concentration of a substance is affected by

    - the ability of the substance to diffuse through the membrane,
    - the overall size and shape of the cell, and
    - the maximum amount of the substance a given area of membrane can transport per unit time.

    a. How do surface-area-to-volume (*SA/V*) ratios change as the size and shape of cells and organisms change? To answer this, calculate the *SA* and *V* of a cube that is 1 mm on a side. Then do the same for cubes that are 2 mm and 4 mm on a side, and compare their *SA/V* ratios.

Cubes:	1 mm square	2 mm square	4 mm square
Linear dimension	1 mm	2 mm	4 mm
Surface area (*SA*)	$6(1 \text{ mm}^2) = 6 \text{ mm}^2$		
Volume (*V*)	$(1 \text{ mm})^3$		
SA/V ratio	$6 \text{ mm}^2/1 \text{ mm}^3$		

b. In general, how does surface area change as linear dimensions increase twofold?

c. In general, how does volume change as linear dimensions increase twofold?

d. In general, how do *SA/V* ratios change as linear dimensions increase twofold?

2. Assume a bacterium is 10 μm in linear dimension. Fill in the chart.

    a. If modeled as a cube, what would its *SA, V,* and *SA/V* ratio be?
    b. If modeled as a sphere, what would its *SA, V,* and *SA/V* ratio be?
    c. What are the *SA* and *V* values and the *SA/V* ratios for a cube-shaped eukaryotic cell that is 100 μm in linear dimension?

	a. 10-μm bacterium as a cube	b. 10-μm bacterium as a sphere	c. 100-μm eukaryote, cube-shaped
SA			
V			
SA/V ratio			

3. Assume that every cell requires a minimum of 1 unit of oxygen per μm^3 per second to stay alive. Fill in the chart.

   a. How much oxygen must cross each μm^2 of surface area per second in the 10-μm bacterium versus the 100-μm eukaryote to keep each alive?
   b. What effects might this difference have on metabolic rates in these organisms?

	10-μm bacterium	100-μm eukaryote
a. Oxygen/μm^2 of *SA*/ second		
b. Possible effect(s) on metabolic rate		

4. Given what you know about cell membranes, is there likely to be a maximum upper limit on the number of molecules of a substance that can cross a given area of membrane per unit time? If so, what factors would be involved in determining the maximum upper limit?

As you answer questions 5, 6, and 7, fill in the chart on the next page.

5. (a) On average, how large is a prokaryotic genome? (b) On average, how many times larger is a eukaryotic genome? (c) Are these genomes haploid or diploid? (*Hint:* See Section 21.2 on pages 393–395 in *Biological Science*, 5th edition.)

6. Under ideal conditions, many bacterial cells can reproduce or duplicate themselves within an hour. Some species (for example, *E. coli*) can reproduce every 20 minutes under ideal conditions. If you innoculate 1 liter of culture medium with one bacterium per milliliter (of medium) at time $t = 0$, how many bacteria will be present in 1 milliliter of the culture medium after 10 hours?

7. Assume that the mutation rate of bacterial cells in culture is 10^{-6} to 10^{-8}. This means that in every 1 million to 100 million cells produced from a single original cell, you would expect to find at least one mutation. How many mutations would you be likely to find in 1 liter of the 10-hour culture of the cells you grew in question 6?

	Prokaryote	Eukaryote
Genome size		
Haploid versus diploid		
Maximum reproduction rate		
Mutation rate per hour		

8. Using all of the information in this activity, write an argument entitled: "How small size in prokaryotes played a role in limiting their morphological diversity and promoting their biochemical (metabolic) diversity."

Name _____   Course/Section _____

Date _____   Professor/TA _____

Activity 30.1 How has endosymbiosis contributed to the diversity of organisms on Earth today?

1. The protists have been called a "catch-all group." What does this mean? Explain.

2. There are more than 100,000 recognized species of protists. Even though this group is so diverse, we can recognize some common themes in their evolution, including the evolution of complex cell structure, novel genetic recombination strategies, and complex life cycles. In addition, this group as a whole has great ecological importance. Provide at least two examples of protists from Chapter 30 in *Biological Science,* 5th edition, that illustrate each theme.

Theme	Examples
a. Complex cell structure	
b. Novel genetic recombination strategies	
c. Complex life cycles	
d. Ecological importance	

3. In the late 1960s (and since), Lynn Margulis provided considerable evidence for the endosymbiotic theory of the origin of various organelles in eukaryotic cells.

a. What is the endosymbiotic theory?	
b. Which two eukaryotic organelles were proposed to have arisen as endosymbionts?	c. What evidence did Margulis present to support each organelle as an endosymbiont?

4. The chloroplasts of the Discicristata, Chromalveolates, and Rhizaria are all thought to have arisen as a result of secondary endosymbiosis.

a. What is secondary endosymbiosis?

b. What evidence supports this idea of secondary endosymbiosis?

5. Structurally, *Giardia lamblia* lacks complete mitochondria and was once thought to be an example of what the earliest living eukaryotes may have looked like.

a. If *Giardia* is similar in structure to the earliest living eukaryotes, what does this imply about the order of evolution of the various eukaryotic organelles (that is, nucleus, cytoskeleton, mitochondria, chloroplasts)?

b. We have since discovered that the diplomonads (such as *Giardia*) and the parabasalids (the other group of excavates) have modified mitochondria.
   • What types of mitochondria do they have?
   • What does this evidence indicate about the evolution of the various eukaryotic organelles?

6. The metabolic pathways of organisms living today evolved over a long period of time—undoubtedly in a stepwise fashion because of their complexity. Considering everything you have learned to date about the evolution of life on Earth, put numbers 1–5 in the following blanks to indicate the order in which these organisms might have evolved, and explain your arrangement. (Note: Some steps happen at the same time and will be part of the same stage.)

___ prokaryotes capable of performing the Krebs cycle

___ eukaryotes capable of performing the Krebs cycle

___ prokaryotes capable of performing electron transport

___ eukaryotes capable of performing electron transport

___ prokaryotes capable of performing glycolysis

___ prokaryotes capable of performing photosynthesis

___ eukaryotes capable of performing photosynthesis

Name_____   Course/Section_____

Date_____   Professor/TA_____

Activity 31.1 What major events occurred in the evolution of the plant kingdom?

Construct a concept map that describes the early evolution of plant life on Earth. Be sure to include relationships among all the organisms and factors in the list on the next page. Keep in mind that there are many ways to construct a concept map. Use these guidelines:

- Begin by writing each term on a separate Post-it note or piece of paper.
- Then organize the terms into a map that indicates how the terms are associated or related.
- Draw lines between terms, and add action phrases to the lines that indicate how the terms are related.

Here is an example:

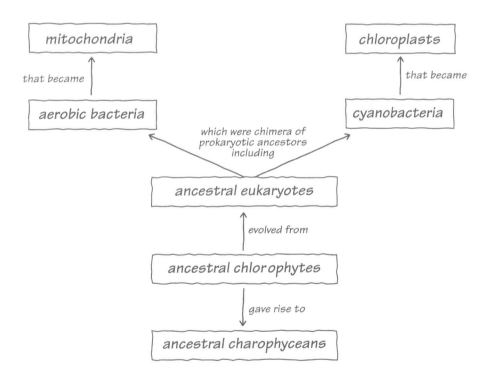

If you are doing this activity in small groups in class, explain your map to another group of students when you finish it.

Terms

bryophytes	microspore	gametangia
pteridophytes	nonvascular plants	root
lycophytes	seedless	stem
endosymbiont	seeds	flagellated sperm
cyanobacteria	angiosperm	archegonium
chloroplast	gymnosperm	antheridium
mitochondria	flowers	pollen grain
vascular tissue	xylem	sporopollenin
waxy cuticle	phloem	sporangia
coleochaetophyceae	spore	stomata
charaphyceae	gametophyte	lignin
alternation of generations	sporophyte	
megaspore	egg	

Use the understanding you gained from constructing the concept map to answer the following questions.

1. Describe the major problems ancestral land plants had to overcome before they could make the transition from water to land.

2. Describe the major solutions to the problems in question 1 that can be found in today's land plants. In other words, what mutations occurred that allowed organisms to make the transition to land?

Activity 31.1

Name _____   Course/Section _____

Date _____   Professor/TA _____

   Activity 31.2 What can a study of extant species tell us about the evolution of form and function in the plant kingdom?

Fill in the chart on these two pages to compare the major features of key groups of land plants with one another and with the green algae.

Plant Group						
	Bryophytes	Lycophytes	Pterophytes	Gymnosperms	Angiosperms	Green Algae
Chlorophyll *a* and *b*	Yes	Yes	Yes	Yes	Yes	Yes
Jacketed gametangia	Yes	Yes	Yes	Yes	Yes	No
Cuticle						
Stomata						
Vascular tissue						
Stems with vascular tissue						
Roots or rhizomes						
True leaves (contain vascular tissue)						
Antheridia						
Archegonia						

(Continued)

Plant Group						
	Bryophytes	Lycophytes	Pterophytes	Gymnosperms	Angiosperms	Green Algae
Flagellated sperm						
Pollen						
Seed						
Flower						
Fruit						
Gametophyte dominant						
Sporophyte dominant						
Sporophyte dependent on gametophyte for energy						
Sporophyte and gametophyte independent						
Gametophyte dependent on sporophyte						

Use the information in the chart to answer the following questions.

1. Some of the major plant groups are listed in the following chart from most primitive to most advanced. For each group, indicate what major characteristics make it more advanced than the preceding group. For example, how are ferns more advanced than mosses?

	Plant Group						
	Green Algae	Bryophytes	Lycophytes	Pterophytes	Gymnosperms	Angiosperms	
Advance(s) over preceding group							

2. How do the bryophytes differ from the seedless vascular plants? How are they similar?

3. The life cycle of all land plants includes an alternation of generations between a multicellular gametophyte phase and a multicellular sporophyte phase. Diagram the life cycle of a seed plant.

a. What cellular division process always precedes formation of the gametophyte generation?

b. What cellular process always precedes formation of the sporophyte generation?

c. If the sexual generation gives rise to the gametes, what part of an angiosperm is sexual?

d. If the sexual generation gives rise to the gametes, what part of a bryophyte moss is sexual?

4. Until the evolution of the seed plants, land plants were dependent on the availability of water for reproduction. Explain why this was true. Explain how seed plants overcame the need for water in reproduction.

5. Pollen, seeds, flowers, and fruits are considered among the most advanced characteristics in the plant kingdom. What evolutionary advantage(s) does each of these offer (relative to what existed before)?

Name _____   Course/Section _____

Date _____   Professor/TA _____

 Activity 31.3 How are the events in plant evolution related?

1. Working in groups of three or four, assign each student in the group one of the
    following events in plant evolution to research. Then give each student 5 minutes to
    report the results of his or her review to the other members of the group.

Events in Plant Evolution:

I. Evolution of vascular tissue
    Mutations in some land plants gave rise to vascular tissue. What advantages did
    these plants have compared with land plants that did not contain any vascular tissue?

II. Evolution of roots and leaves
    Mutations in some land plants gave rise to roots, leaves, or both. What advantages
    did these plants have compared with plants that did not contain roots or leaves?

III. The trend toward reduction of the gametophyte generation
    Mutations in some land plants gave rise to life cycles in which the gametophyte
    generation was reduced. What advantages did these plants have compared with
    plants that did not have a reduced gametophyte generation?

IV. Evolution of the seed
    Mutations in some land plants gave rise to the seed. What advantages did these
    plants have compared with plants that did not have seeds?

2. Work together as a group to determine how these events in plant evolution (I to IV) might be related. For example, which would have to come first (in evolution), which next, and so on? Another way to look at this question is to consider which of these events paved the way (or made it possible) for the other events to occur.

   Be sure to state evidence for your proposed evolutionary scheme. To do this, it is useful to ask yourself questions like these: Would it have been possible for the seed to evolve without vascular tissue first having evolved? Leaves? And so on. If yes, how could this have occurred? If no, why not?

3. Incorporate the following *true* observations into your analysis of how the events in plant evolution (I to IV) could be related:

   a. The fossil record (spore evidence and so on) indicates that the first plants on Earth were similar to modern-day bryophytes.

   b. The very large marine brown algae (for example, giant kelp) can grow to lengths of 30 feet or more. These algae have both leaf-like and stem-like structures and are held to the bottom of the sea by a holdfast. When examined microscopically, the algae are found to contain transport vessels that are similar to phloem in function. These giant kelp do not contain any xylem-like vessels, however, nor do they have roots.

   c. The first land plants with xylem and phloem had no leaves or roots.

   d. Some seedless vascular sporophyte plants do not release megaspores. Instead, the megaspore divides on the sporophyte (in the old sporangium) to produce the female gametophyte. This female gametophyte produces eggs in archegonia, which are fertilized by sperm produced in the antheridia of this or other plants. Neither the female gametophyte (once formed) nor the developing embryo receives nutrition from the old sporophyte plant.

4. Write an analysis of how these events could be related.

5 to 8. Consider a puddle of water in which mosses, ferns, and an angiosperm are growing. In a drop of water taken from the puddle, you observed flagellated sperm. Which of the following statements concerning the sperm could be true, and which are definitely false? Consider each statement separately, and explain your answers.

T/F 5. The sperm could have been produced by an antheridium on the moss sporophyte.

T/F 6. The sperm could later stop swimming and develop into the male gametophyte of the fern.

T/F 7. The sperm may be swimming to the archegonium of the fern.

T/F 8. The sperm could have been released by a pollen tube.

Name _____   Course/Section _____

Date _____   Professor/TA _____

Activity 32.1 How diverse are the fungi in form and function?

1. a. What is the basic body plan of most fungi?

    b. Which fungi do not share this basic body plan?

2. a. In what ways are fungi important in the ecosystem?

    b. In what ways are fungi important to humans?

3. Fungi are much more closely related to animals than they are to land plants (see Figure 32.7 on page 618 in *Biological Science,* 5th edition). In addition to the DNA sequence data, several morphological traits link animals and fungi. What are some of these animal-like characteristics?

4.

a. Into what five major phyla (or divisions) is the kingdom Fungi divided?	b. On what basis are these divisions made?

5. Is it more correct to describe a mushroom as haploid or diploid? Explain.

6. If you did not know that fungi were primarily terrestrial organisms, what structures or features of the organisms would suggest that they were terrestrial?

7. Lichens are symbiotic associations of a fungus, usually an ascomycete, and an alga, usually a green algal species or a cyanobacterial species. Lichens can often survive in harsh natural environments.

   a. To what environmental conditions are lichens well adapted?

   b. What makes them so well adapted to these conditions?

Name _____   Course/Section _____

Date _____   Professor/TA _____

Extend Your
Understanding   **Activity 32.2 Can fungi be autotrophic or photosynthetic?**

Recently, Ekaterina Dadachova and colleagues found that the fungi *Cryptococcus neoformans, Wangiella dermatitidis,* and *Cladosporium sphaerospermum* use melanin, also a pigment found in human skin, to transform radiation into energy to use as food for growth. Researchers believe that melanin is present to protect fungi from stress, such as radiation, and that certain species use this molecule for metabolic reactions. *Cryptococcus neoformans* is currently in the phylum Basidiomycota, and both *W. dermatitidis* and *C. sphaerospermum* are classified as ascomycetes. If this discovery is substantiated with further research, these and possibly other species of fungi may be reclassified as autotrophic or even photosynthetic. Given this new information about these species of fungi, where would you place them on the fungal phylogeny? Draw where you would put each species on the tree below, and justify your answer. Feel free to do additional research on each species to help inform your answer, and remember to cite your sources.

Name _____   Course/Section _____

Date _____   Professor/TA _____

Integrate Your
Understanding   **Chapters 33–34(1)—What can we learn about the evolution of the animal kingdom by examining modern invertebrates?**

Fill in the chart on the next two pages to organize the major characteristics of key invertebrate phyla.

Key invertebrate phyla

Characteristics	Porifera	Cnidaria	Platyhelminthes	Nematoda	Annelida	Mollusca	Arthropoda	Echinodermata
Examples of organisms	Sponges							
Number of tissue layers in embryo	Doesn't apply	2; ectoderm and endoderm						
Tissue versus organ level development	Quasi tissue level	Tissue level						
True muscle	No	No; have epithelio muscular cells						
Symmetry? Cephalization?								

Key invertebrate phyla

Characteristics	Porifera	Cnidaria	Platyhelminthes	Nematoda	Annelida	Mollusca	Arthropoda	Echinodermata
Coelom? Type?								
Digestive tract? Type?								
Circulatory system? Type?								
Nervous system? Type?								
Other								

Using the information in the chart and in Chapters 33 and 34 of *Biological Science*, 5th edition, answer the following questions.

1. What set of characteristics is shared by all of the invertebrate animal phyla in the chart?

2. What unique combination of characteristics defines each of the invertebrate phyla as separate from the other phyla?

3. If you compare the characteristics of one phylum of the invertebrates with the next, what key differences separate the groups from each other?

4. a. Looking across the rows, what major trends appear to occur in the evolution of various organs or organ systems in these animal groups?

    b. Does this analysis provide evidence for or against the statement: "Evolution adds onto or modifies what already exists"? Explain.

5. The chart organizes the major groups of animals based on grade or shared body plan features. What changes would you need to make in this organization to reflect the possible phylogenetic relationships uncovered using molecular evidence?

6. Certain developmental characteristics are used to determine evolutionary relationships among animal phyla.

   a. What developmental evidence was used to link annelids, arthropods, and mollusks evolutionarily?

   b. What evidence is used to separate these phyla from the echinoderms and chordates?

7. In biological terms, a group of organisms is said to be successful if represented by a large number of species, or if the mass of all the organisms in the group is large. (In both cases, *large* is determined relative to other groups or organisms.) Given this definition of success, which of the major groups of animals would you argue is the most successful? Be sure to provide evidence for your argument.

Name _____   Course/Section _____

Date _____   Professor/TA _____

Integrate Your
Understanding **Chapters 33–34(2)—What factors affect evolution**
of organisms as they become larger?

In the evolution of life on Earth, organisms have evolved from single-celled to multicelled; from small to large; from simpler to more complex. Keep in mind that this apparent increase in complexity occurs because evolution adds on to or modifies what already exists.

As you discovered in Activity 29.2, surface-area-to-volume ratios and the need for organisms to gain and/or lose substances across their surface areas have put constraints on evolution of cell structure and function. These same constraints affect the ability of multicellular organisms to survive. Given the small size of unicellular organisms, most of their evolution has involved changes in cell chemistry and/or internal cellular structure. The opposite is true for evolution of larger and larger multicellular organisms, where evolution is evidenced primarily as changes in both external and internal morphology.

1. A quick review of Activity 29.2 (How has small size affected prokaryotic diversity?): How do surface-area-to-volume (SA/V) ratios change as the size and shape of cells and organisms change? To answer this, calculate the SA and V of a cube 1 mm on a side. Then do the same for cubes that are 2 mm and 4 mm on a side, and compare their SA/V ratios.

Cubes:	1 mm square	2 mm square	4 mm square	Ratios
Linear dimension	1 mm	2 mm	4 mm	1:2:4
Surface area (SA)	$6(1 \text{ mm}^2) = 6 \text{ mm}^2$			
Volume (V)	$(1 \text{ mm})^3$			
SA/V ratio	$6 \text{ mm}^2/1 \text{ mm}^3$			

a. What do the ratios mean?

b. Is this graph of the change (in surface area compared to volume) linear or exponential? To answer this, complete the following table and graph the surface areas and volumes from this table and the table above.

Cube linear dimension:	8 mm	16 mm	32 mm	64 mm
Surface area (mm^2)				
Volume (mm^3)				

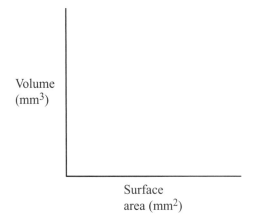

c. What effect does this relationship between surface area and volume have on the ability of larger and larger multicellular organisms to support the metabolic needs of all parts of their bodies, for example, to supply the oxygen needed for cell respiration?

2. In the previous examples, as the "organism" sizes changed, their shapes remained constant (as cubes). How would changing the shape of the organism affect *SA/V* ratios? Use modeling clay or playdough to make a cube that is 2.5 cm (1 inch) on a side. The cube will obviously have a constant mass or volume.

Using your model, devise three ways to change the *SA/V* ratio of this "organism." If possible, find one way to reduce the *SA/V* ratio and two ways to increase the *SA/V* ratio. For each of your solutions to increase or decrease the *SA/V* ratio, calculate the actual change in *SA*. Enter the data you collect and/or calculate in the chart.

	Shape change	Effect on *SA/V* ratio
Model 1		
Model 2		
Model 3		

3. Review the chapters on fungal, animal, and plant diversity in your textbook. For each of the models that you developed to increase *SA/V* ratios, find an example of a fungus, plant, and/or animal that uses the same method to increase *SA/V* ratios and describe it below.

Proposed model—Increase *SA/V* by:	This type of increase in *SA/V* ratios can be found in the following organisms and/or organ systems:

Name _____   Course/Section _____

Date _____   Professor/TA _____

  Activity 35.1 What can we learn about the evolution of the chordates by examining modern chordates?

Fill in the chart on the next two pages to organize the major characteristics of key groups of chordates.

Characteristics of phylum chordata

Characteristics	Sub phylum Urochordata	Subphylum Cephalochordata	Subphylum Vertabrata							
			Craniata	Chondrichthyes	Actinopterygii	Sarcoperygii	Amphibia	Mammalia	Reptilia	Aves
Examples of organisms										
Type of skeleton										
Vertebral column?										
Jaw?										
Paired limbs? Stricture? Limb position?										
Body covering?										

Characteristics of phylum chordata

Characteristics	Subphylum Urochordata	Subphylum Cephalochordata	Subphylum Vertabrata								
			Craniata	Chondrichthyes	Actinopterygii	Sarcopterygii	Amphibia	Mammalia	Reptilia	Aves
Respiratory structure?										
Circulatory system? Heart structure?										
Type of reproduction										
Other										

Name _____   Course/Section _____

Using the information in the chart, answer these questions.

1. a. According to *Biological Science,* 5th edition, what three characteristics are common to all chordates?

    b. Why don't we find all three of these chordate characteristics in our own bodies?

2. The chart organizes the chordates based on grade or shared body plan features. If you compare the characteristics of one group with the next, what key differences separate the groups from each other?

3. What unique combination of characteristics defines each group as separate from the others?

4. a. Looking across each of the rows in the chart, what major trends do you see in the evolution of the different organs and organ systems?

   b. Does this analysis provide evidence for or against the statement: "Evolution adds onto or modifies what already exists"? Explain.

5. a. What major changes in structure and function are seen in terrestrial groups as compared with aquatic groups?

   b. Can these changes be related to differences in natural selection on land versus those in an aquatic environment? Explain.

Name _____   Course/Section _____

Date _____   Professor/TA _____

Chapters 29–35 How do organisms interact with each other and their environments?

For each major grouping of organisms (Bacteria, Archaea, Protists, Green Plants, Fungi, Protostomes, and Deuterostomes) detailed in Chapters 29–35 of *Biological Science,* 5th edition, select two organisms or key lineages. Using these 14 organisms/groups of organisms, along with the N and C cycles shown in Chapters 29, 30, and 32, write a biologically accurate short story about the interactions between organisms and between organisms and their environments.

Name _____   Course/Section _____

Date _____   Professor/TA _____

Activity 36.1 How do viruses affect host cells?

1. By definition, viruses are obligate intracellular parasites. What does this mean?

2. In general, how are viruses classified?

3. What is reverse transcriptase? Where was reverse transcriptase first found? Of what use is it to viruses? Of what use is it to scientists?

4. Why must viruses invade other cells to reproduce?

    a. Describe the general process of virus reproduction. Include a discussion of lytic and lysogenic viral cycles.

    b. Which types of viruses are more likely to have a lysogenic phase?

5. More than 100 different viruses—many of them rhinoviruses—can cause the "common cold" in humans. Other viruses, influenza viruses, cause the flu. Although there are many different antibiotics for treating bacterial infections, relatively few drugs are available to treat viral infections. Explain.

Integrate Your
Understanding **Chapters 29, 32, and 36—Develop a Drug Combination to Effectively Combat Multiple Infections**

Patients with AIDS often acquire opportunistic infections. Imagine a patient with AIDS and two opportunistic infections—a respiratory disease caused by a *Mycobacterium* and another disease caused by the fungus *Candida*. As the patient's physician, you need to prescribe drugs to counteract the infections.

Given the normal anatomical and functional characteristics of each disease organism, what characteristics should a drug combination have to treat each of these three infections while doing the least harm to the patient?

Name _____   Course/Section _____

Date _____   Professor/TA _____

Activity 37.1 How does plant structure differ among monocots, herbaceous dicots, and woody dicots?

The chart below shows a drawing of a generic plant. The arrows indicate various plant parts or organs. In the columns to the right of the plant, draw cross sections of the plant at the points indicated by the arrows.

To help you visualize the differences in structure among different types of plants, in Column 1, draw the cross sections assuming the plant is a monocot. In Column 2, draw the cross sections assuming the plant is an herbaceous dicot. Finally, in Column 3, draw the cross sections as if the plant is a woody dicot. Be sure to label your drawings.

Generic plant	Column 1 Cross sections if plant is a monocot	Column 2 Cross sections if plant is an herbaceous dicot	Column 3 Cross sections if plant is a woody dicot
a b c d Soil surface e f			

Use the information in your drawings to answer the following questions.

1. In Column 1, connect the cross sections of the monocot by drawing a line (color 1) from the water transport tissue in one section to the water transport tissue in the next, and so on. Draw another line (color 2) to connect the food transport tissue from one section to the next. Do the same for the herbaceous and the woody dicot cross sections.

2. Use the information in the cross sections to fill in the chart below. *Note:* A distinguishing feature is one that is found in only the given type of organism. For example, a distinguishing feature of mammals is the presence of hair: Hair is a characteristic of all mammals and is not found in any other animals.

| | Distinguishing feature(s) of | | | |
	Leaf	Stem	Root	Apical meristem shoot vs. root
Monocot				
Herbaceous dicot				
Woody dicot				

3. If cross sections were not available, what other characteristics of the plants as a whole could you use to determine whether each was a monocot, an herbaceous dicot, or a woody dicot?

4. A cartoon shows a man going to sleep in a hammock suspended between two relatively short trees. The second frame of the cartoon shows the man waking 20 years later and finding his hammock 15 feet higher off the ground. Critique this cartoon in terms of what you know about the growth pattern of trees.

Name _____   Course/Section _____

Date _____   Professor/TA _____

Extend Your
Understanding   **Activity 37.2 How are the structure and function of plant organs correlated?**

How is the general morphology of the various organs of a plant correlated with the function(s) of those organs? For example, why are leaves generally thin and flat? What structural advantage is provided to stems that have their vascular tissue arranged in a ring near the periphery of the stem? What characteristics of roots and root growth dictate that the vascular tissue be more centrally located?

Name _____   Course/Section _____

Date _____   Professor/TA _____

  Activity 38.1 How are water and food transported in plants?

Refer to the cross sections of the root you developed in Activity 37.1, and trace the pathway of a water molecule from the soil until it leaves the leaf at the top of a tree. Then use modeling clay, cut-out pieces of paper, chalk, or some other material to create a model of transport. Be sure to include all the terms and concepts in the list below.

- Model or demonstrate how water is transported from the soil to the xylem.
- Model or demonstrate the transport of K^+ ions from the soil to the xylem.

Terms

endodermis	symplastic transport
oxidative phosphorylation	ATP
Casparian strip	apoplastic transport
oxygen	epidermis
plasmodesmata	symplast
respiration	lateral transport
stele	apoplast
channels	absorption
root hairs	xylem
ADP	$-\Psi$ versus $+\Psi$

Use your model to answer the following questions.

1. How do water's properties of adhesion and cohesion help maintain the flow of water in the xylem of a plant?

2. a. If water flows from a region of more positive (higher) water potential to a region of more negative (lower) water potential, how does the water potential in the root compare to that in the soil outside the root?

    b. How does the water potential in the air compare to that in the leaf of a plant undergoing transpiration?

3. A student uses an $^AU^B$ tube for a series of experiments. Sides A and B of the tube are separated by a membrane that is permeable to water but not to sugar or starch. What results would you expect under the experimental conditions given below? Explain your answers in terms of osmotic potential, water potential, and the equation

$$\Psi = \Psi_P + \Psi_s$$

(*Hint:* Solute pressure is always negative, and a 0.1 *M* solution of any substance has $\Psi_s = -0.23$. Therefore, a 0.2 *M* solution would have $\Psi_s = -0.23 \times 2 = -0.46$.)

Experiment a: A solution of 10 g of sucrose in 1000 g of water (the molecular weight of sucrose is 342) is added to side A. An equal volume of pure water is added to side B. What will happen to the concentrations of water and sugar in the two sides over time? Explain.

Experiment b: A solution of 10 g of soluble starch in 1000 g of water is added to side A. Assume the molecular weight of soluble starch is about 63,000. An equal volume of pure water is added to side B. What will happen to the concentrations of water and starch in the two sides over time? Explain how this compares with the results in Experiment a.

4. Fertilizer generally contains nitrogen and phosphorus compounds required by plants. The nitrogen is often in the form of nitrates, and the phosphorus is in the form of phosphates. Based on what you know about chemistry and water potential, why would overfertilizing lead to the death of plants?

5. a. One of the most common ways of killing a plant is overwatering. Why does overwatering kill a plant?

   b. If overwatering kills plants, why can you sprout roots from cuttings of stems in water?

6. Xylem cells are dead when functional. Why must phloem cells be alive when functional?

7.

What forces bring about:	How are these forces generated in the plant?
a. xylem conduction?	
b. phloem conduction?	

8. Refer to your diagram of the cross-sectional structure of a typical angiosperm leaf from Activity 37.1. Correlate this structure (that is, the type and placement of cells) with the activities of the leaf as they relate to photosynthesis, water conservation, and food and water transport.

Name _____   Course/Section _____

Date _____   Professor/TA _____

 Extend Your Understanding   **Activity 38.2 Plant Transport Experiments**

1. Scientists have measured the circumference of trees at 2 a.m. and at 2 p.m. If they collect measurements when the ground has adequate moisture and the days are sunny and dry, they find that the circumference (and therefore the diameter) of the tree trunks is smaller at 2 p.m. than at 2 a.m. From your knowledge of the mechanisms of water transport, suggest the reasons for this decrease in circumference.

2. Outline an experiment that would allow you to determine: (a) how fast a substance is carried in the xylem, (b) in what direction the substance flows in the xylem, and (c) what percentages of solutes are in the xylem at various distances away from the leaves or roots.

3. When researchers have tried to tap into phloem cells during experiments, they find that the disrupted phloem immediately stops functioning. However, aphids can pierce through plant tissues with their mouthparts and locate individual phloem cells. Once inside a phloem cell, the aphids are essentially force-fed phloem sap. If the aphid body is removed from the mouthparts, phloem sap will continue to flow and can be collected.

   Place two aphids on a plant you are using for radioactive tracer experiments: One should be located 5 cm above leaf A; and the other, 5 cm below leaf A (see diagram on the next page). Remove the aphid bodies so you can sample phloem sap from the mouthparts, cover leaf A with a plastic bag, and inject $^{14}CO_2$ (radioactive carbon dioxide) into the plastic bag. Leaf A is allowed to conduct photosynthesis for 10 minutes, during which time $^{14}CO_2$ is continually supplied to the bag. Finally, collect honeydew from the two aphids' mouthparts at 30-second intervals for one hour, starting from the time the $^{14}CO_2$ is first injected into the bag. If you analyze the phloem samples for ^{14}C, where would you expect to find it? Explain your reasoning.

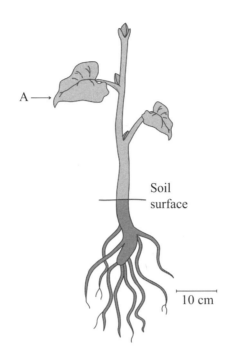

A →

Soil
surface

10 cm

4. Outline an experiment that would allow you to determine:
   (a) how fast a substance is carried in the phloem,
   (b) in what direction the substance flows in the phloem, and
   (c) what percentages of solutes are in the phloem at various distances away from the
       leaves or roots.

Name _____   Course/Section _____

Date _____   Professor/TA _____

 Activity 39.1 What do you need to consider in order to grow plants in space (or anywhere else, for that matter)?

Long-range, manned space travel and manned space stations may someday become a reality. Before this can occur, however, we will have to develop sustainable methods of agriculture suitable for use in space. One of the key methods being investigated is hydroponics—growing plants in water supplemented with nutrients.

You are assigned to a team working on the design of plant growth systems for use in a space station.

1. What types of plants would you choose to grow? Explain the reasoning behind your choices.

2. When you set up the growth system:

a. What nutrients would you need to add to the water? List what you would need, and explain why each nutrient would be necessary.	b. What atmosphere would you need to maintain? List what components you would need to maintain in the atmosphere, and explain why each component would be necessary.	c. Which of the requirements in parts a and b could be recycled? Which of them could not be recycled? Explain.

3. You are interested in growing soybeans that have a higher amino acid content than those varieties that are currently available. How would you do it?

4–5. Pine seedlings grown in sterile potting soil grow much slower than seedlings grown in soil from the area where the seeds were collected. Which of the following is a reasonable explanation? Explain your answers.

   A = is a reasonable explanation.
   B = is not a reasonable explanation.

4. The normal symbiotic fungi are not present in the sterilized soil.

5. Water uptake is faster when mycorrhizae are present.

6. You observe the results shown in the graph below by measuring plant growth in two different soils (soil A and soil B).

a. Assuming a difference between the soils is the presence of mycorrhizae in one soil and its absence in the other, which soil has mycorrhizae? How do you know?

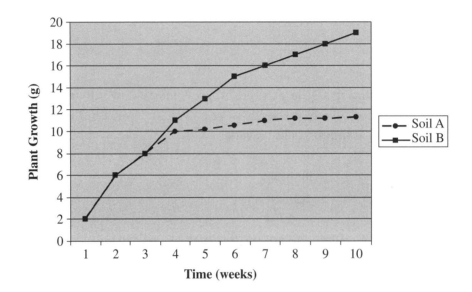

b. Propose an explanation for why the plants in soils A and B grow at nearly the same rate for the first few weeks.

Activity 39.1

Activity 40.1 How do gravity and light affect plant growth responses?

Review Chapter 40 of *Biological Science,* 5th edition, and your answers to Activity 39.1. Then answer the following questions.

---

1. One of the problems associated with growing plants in space is the lack of gravity.

---

a. How does gravity affect the normal growth of a plant's roots, stems, and other parts? Explain the mechanisms involved.

---

b. How would a lack of gravity affect normal plant growth?

---

c. Propose mechanisms to overcome the problems associated with a lack of gravity.

---

2. Another problem with growing plants in space relates to a plant's light requirements and phototropic responses versus the photoperiods required for the plant to flower and produce fruit.

   a. How do phototropism and photoperiodism differ?

   b. What light characteristics would you use to maximize plant growth per unit time?

   c. What kind of physical environment would you need to maintain appropriate photo-tropic responses among plants?

   d. What design modifications would you need to make to support plants with different photoperiods—for example, long-day versus short-day plants?

Extend Your
Understanding **Activity 40.2 Plant Response to Signals: Sample**
Exam Questions

1. Compared to the wild type, a mutant plant that overproduces cytokinin will probably have _____ lateral branches. (Fill in the blank with the best answer.) Explain your answer.
   A. more
   B. fewer
   C. the same number of

2. Combining the mutation in question 1 above with one that causes the overproduction of auxin would _____ the effect you chose above. (Fill in the blank with the best answer.) Explain your answer.
   A. increase
   B. decrease
   C. not change

Questions 3 to 4. A species of long-day plant will flower when the days are over 14 hours long. Which of the following light cycles would cause this plant species to flower?

Answer *true* if the cycle causes flowering and *false* if it does not. Explain your answers.

T/F   3.   Continuous white light day and night, interrupted every 4 hours by flashes of red light.

T/F   4.   A daily cycle of 9 hours of light followed by three periods of 4 hours and 50 minutes of uninterrupted hours of dark and 10 minutes of red or white light each.

Copyright © 2014 Pearson Education, Inc.

Name _____   Course/Section _____

Date _____   Professor/TA _____

Activity 41.1 How can plant reproduction be modified using biotechnology?

Answer questions 1–7. Then, in question 8, you will develop a concept map that links the information from all seven answers.

1. Draw a general diagram of the life cycle of a seed plant. Indicate which steps are haploid and which are diploid.	2. Define *microsporogenesis* and *megasporogenesis*. In what portion(s) of the flower does each of these processes occur? What is the end-product of each process?

3. Draw and label all parts of a complete flower. Indicate the functions of the major parts.	4. What is pollination? How does it differ from fertilization?	5. Draw and label a mature ovule. Include the micropyle, polar nuclei, egg, and antipodals. Indicate the functions of each of these structures.
	6. What stages of the life cycle are eliminated or bypassed when plants are cloned naturally? When plants are cloned on the farm or in the laboratory?	7. What does the science of plant biotechnology do that artificial selection and/or cloning practices do not do?

8. Construct a concept map that links all of the information in questions 1–7. You can do this in the space below.

Name _____ Course/Section _____

Date _____ Professor/TA _____

 **Activity 42.1 How does an organism's structure help
it maintain homeostasis?**

1. To remain alive, an organism must be able to maintain homeostasis of its internal
    environment relative to the external environment. What structure(s) or system(s)
    are of primary importance in maintaining homeostasis in the following situations
    in amoeba versus mammal?

Situation	Amoeba	Mammal
a. Change in environ-mental pH and/or temperature		
b. Reception of stimuli, for example: light and touch		
c. Response to stimuli		

2. Cells must be bathed continuously in an aqueous medium to take in oxygen and nutrients and get rid of waste products via diffusion. Diffusion is efficient over only short distances. In fact, diffusion is efficient only for a maximum distance of about three cell diameters (approx. 200 to 300 µm). Note the following times required to diffuse specific distances:

Diffusion distance (µm)	Time required for diffusion
1	0.5 msec
10	50 msec
100	5000 msec (5 sec)
1000 (1 mm)	500,000 msec (8.3 min)
10,000 (1 cm)	50,000,000 msec (14 hr)

a. Graph the data in the table, and calculate the mathematical relationship between the increase in distance and the corresponding increase in time required to diffuse that distance. You can do this by hand, or you can use a software program with graphing capability. Include your graph and the results of your calculations below.

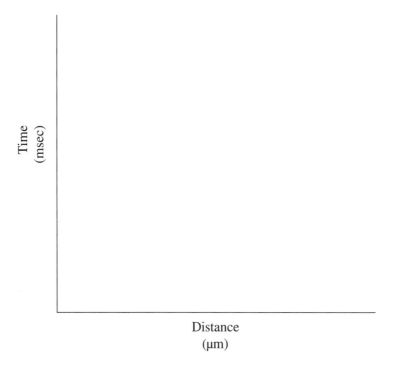

Distance
(µm)

b. How much time would be required for oxygen to diffuse 5 µm? 200 µm?

Activity 42.1

Name _____   Course/Section _____

3. What happens to the surface-area-to-volume (*SA/V*) ratio of a three-dimensional object (such as a cell) as its linear dimension increases? For example, how does the *SA/V* ratio of a sphere or cube change as the linear dimensions increase? (Formulas for a sphere: surface area $= 4\pi r^2$; volume $= {}^4/_3\pi r^3$.) (Also see Activity 7.1.)

4. What effect(s) are the physical properties of diffusion likely to have on the design of epithelia and epithelial cells and digestive and circulatory systems as animals become larger and larger? In your answer, consider also how *SA/V* ratios change as organisms become larger and the effects this change is likely to impose on the design of organisms.

Activity 43.1 What is nitrogenous waste, and how is it removed from the body?

In the space below, draw a longitudinal section of a mammalian kidney. Next to this, draw a blowup of a nephron (including Bowman's capsule and the loop of Henle) and its associated collecting duct. Be sure to include the vasa recta, glomerulus, and renal corpuscle that are associated with this nephron.

You may do your drawing in chalk on a tabletop or blackboard if they are available.

Use your drawing and your understanding of the operation of the kidney to answer the following questions.

1. Define excretion, and indicate how it differs from elimination.

2. The removal of nitrogenous wastes (excess nitrogen) is a special problem in most animals.

    a. Where does the nitrogenous waste come from?

    b. What is it about the chemistry of nitrogen that makes it difficult for most animals to deal with?

3. Work through parts a and b, and then use the information you gather there to answer the question in part c.

    a. Describe the composition of the newly filtered solution that enters Bowman's capsule. Then compare it to the composition of the blood entering and leaving the glomerulus.

    b. Starting with the solution that escapes into Bowman's capsule from the glomerulus, describe the changes that occur in its composition as it moves through each of these regions:

       i. Proximal convoluted tubule

       ii. Loop of Henle

       iii. Distal convoluted tubule

       iv. Collecting duct

       v. Urinary bladder

c. Now explain how the general function of the kidney enables it to "remove" (a better expression would be "let out") from the body a wide variety of unfamiliar substances (drugs, inorganic molecules, or ions of many kinds) that the body has never encountered before. After answering this, explain why "let out" from the body is a better expression than "remove."

4. It is useful to consider the excretory system (along with the digestive and gas-exchange systems) as primarily involved in bulk exchange with the external environment. The excretory system could also be interpreted as a specialized part of the external surface of the organism, which in its own way encloses and modifies part of the environment. Describe how this is true for the human kidney. For example:

a. Where in the kidney does the organism end and the environment begin?

b. Are changes in the glomerular filtrate changes in the organism, changes in the environment, or both?

c. What do your answers in parts a and b indicate about the possible evolutionary origins of the kidney?

5. a. What is the difference between hydrostatic pressure and osmotic pressure?

b. Where in the human excretory system is hydrostatic pressure responsible for moving water across a membrane or layer of cells?

c. In the excretory system, what is the function of osmotic pressure?

Name _____   Course/Section _____

Date _____   Professor/TA _____

Extend Your
Understanding   **Activity 43.2 Osmoregulation: Sample Exam Questions**

1. a. You examine the kidney structure and function of two species of mice, one from the desert and another from a meadow or grassland. What differences would you expect to find? Explain your reasoning.

    b. Would you expect the excretory systems of organisms that live in the sea to more closely resemble those of animals that live in deserts or those that live in freshwater? Explain.

2. Many medicines are taken orally—that is, swallowed and absorbed from the digestive system. Aspirin and many antibiotics are examples. Even though these drugs may be very different chemically, the instructions for taking them often say, "Repeat dose every 4 to 6 hours." What is the reasoning behind this dose rate? In other words, why do you need to take the medicine every 4 to 6 hours?

Indicate whether the following statements are true or false. Explain your reasoning.

Questions 3 to 5. In a normal, healthy human, urine:

T/F   3. includes nitrogenous waste products formed following deamination of amino acids

T/F   4. functions to rid mammals of excess glucose

T/F   5. production involves a countercurrent principle of exchange

Questions 6 to 8. Injection of an individual with antidiuretic hormone (ADH) would cause:

T/F   6. increased urine volume

T/F   7. decreased filtration at the glomerulus

T/F   8. an increase in the permeability of the collecting ducts

Name _____   Course/Section _____

Date _____   Professor/TA _____

 **Activity 44.1 How are form and function related
in the digestive system?**

1. What is the overall function of digestion (as a whole)?

2.

a. How do bacteria eat?	b. How do amoebas eat?

(Continued)

How are their eating styles similar to what we see in humans? How are they different?			
c. Similarities with bacteria	d. Differences from bacteria	c. Similarities with amoebas	d. Differences from amoebas

3. Label the parts of the mammalian digestive tract in the diagram, and state the major function(s) of each part.

Name _____   Course/Section _____

How does the structure of each part reflect its function?

a. Mouth/teeth	b. Esophagus	c. Stomach	d. Small intestine	e. Large intestine

4. Make a larger copy of the diagram in question 3. Use this larger diagram to model or trace what happens to a food particle from the time it enters the mouth until its indigestible remains are egested, or eliminated. Include all of the following terms in your diagram. Also, note on the diagram the function of each and where it is found in the digestive system.

carbohydrate	lipase	gastric acid (HCl)
fat	pepsin	bolus
nucleic acid	salivary amylase	microvilli
protein	bile	lacteals
cardiac orifice	dipeptidase	capillaries
epiglottis	saliva	feces
pyloric sphincter	amylase	intestinal bacteria
pharynx	pancreas	hepatic portal vein
pepsinogen	nuclease	fiber or roughage

Using the understanding of the structure and function of the digestive system you gained from the model in question 4, answer the following questions.

5. The mammalian digestive tract has been called an extension of the outside world that you enclose in your body.

   a. What does this statement mean? Consider what would happen if you swallowed a marble. Is the marble ever "inside your body"?

   b. At what point in the digestive process is food officially inside the body?

   c. How is mammalian digestion more efficient than the type of digestion seen in bacteria?

6. Digestion (in humans and many other animals) is both physical and chemical. Among the chief chemical agents of digestion are the digestive enzymes. What do enzymes do to food?

7. Have you ever heard the old adage: "Be sure to chew your food 20 times before swallowing"?

   a. What, if any, effect would this chewing have on how well the digestive system functions? Keep in mind that enzymes work only on the surfaces of food particles. Explain.

   b. How does the function of the teeth complement the function of one of the digestive chemicals in the stomach?

8. Although enormous quantities of various enzymes are added to the contents of the duodenum of the small intestine, no traces of enzymatic activity are left in the intestinal contents when they pass into the large intestine (colon). Why? What happens to the enzymes?

9. Most of the blood that leaves the digestive tract of a human is collected into a series of veins that merge to form the hepatic portal vein. The hepatic portal vein carries blood to the liver, where the hepatic portal vein breaks down again into a system of liver capillaries. The liver capillaries drain into the hepatic vein, which carries blood to the vena cava. The vena cava carries blood from the body to the right atrium of the heart. Some of the products of digestion enter a different system of transport, the lacteal system. The lacteal system bypasses the liver and carries its contents directly to the right atrium of the heart.

a. Which products of digestion are carried in the blood to the liver?

b. Which products of digestion are carried via the lacteal system?

c. During the first hour after a heavy meal, how does the concentration of glucose in the blood going from the small intestine to the liver compare to the concentration entering the right side of the heart?

d. Similarly, how does the concentration of amino acids compare?

e. How does the concentration of fat leaving the small intestine compare to the concentration in the right atrium?

10. How does a herbivore such as a cow extract the glucose from the cellulose in its diet? What characteristics of the structure and function of the digestive tract of a ruminant suit it for this function?

11. A peanut butter and jelly sandwich contains a variety of carbohydrates, fats, and proteins. Complete the graphs below to indicate the relative percentages of carbohydrate, fat, or protein that remain in this ingested food as it progresses from your mouth through your digestive tract. Explain your reasoning.

Mouth → esophagus → stomach → small intestine → large intestine

Digestive Tract

12. A good rough generalization is that the more meat in the diet of a species of animal, the shorter its intestine. In comparison, herbivores have long intestines (length always being relative to total body length). How can this trait be explained?

13. What would happen to the normal function of the digestive tract if part of one of the following organs was removed or greatly reduced in size (for example, as the result of surgery following an accident)? How would a person's eating habits need to change to accommodate the reduction in size?

    a. Stomach

    b. Small intestine

    c. Large intestine

14. A friend of yours has come up with a new idea for people who don't have time to eat. He has developed a high-powered Blenderizer that breaks food up into very small particles. He has tested his product on amoeba and *Paramecia*. Both of these single-celled organisms can pick up these small particles and incorporate them into food vacuoles. Your friend is hoping to market the Blenderizer to busy people who could blenderize their food and use an IV bag and tube to run the food directly into their blood system. Because people could hide the system under their clothes, they could essentially eat at any time without it interfering with meetings and other activities. Your friend comes to you for advice and possible financial support to get his idea off the ground. What would you say to him?

Name _____   Course/Section_____

Date _____   Professor/TA_____

  Activity 45.1 How do we breathe, and why do we breathe?

This activity is designed to give you an understanding of how and why we breathe and some of the mechanisms that control the rates of breathing.

How do we breathe? Explain the mechanics of breathing. To do this, diagram (or model) and explain the general movement of oxygen and carbon dioxide into and out of the lungs, lung alveoli, blood, and tissues. Be sure to include all of the following terms in your diagram.

Terms

diaphragm	carbon dioxide
diffusion	brain
CO_2 removal	P_{CO_2} sensors
O_2 demand	medulla oblongata
lungs	lung tissue cells
alveoli	skeletal muscle cells
expired air	carbohydrate $+ O_2 \leftrightarrow CO_2 + H_2O +$ energy
inspired air	$CO_2 + H_2O \leftrightarrow H_2CO_3 \leftrightarrow HCO_3^- + H^+$
oxygen	

Use your diagram or model to answer the following questions.

1. Explain briefly why we breathe. Or, more specifically, where is what we breathe in used in the body and where is what we breathe out produced in the body?

2. Oxygen is transported in the blood in reversible combination with hemoglobin.

    a. How does the level of carbon dioxide in the blood contribute to blood pH? (Write the chemical reaction.)

b. How does the level of CO_2 in the blood affect the affinity of hemoglobin for oxygen? That is, how does the concentration of CO_2 in the blood change its ability to carry oxygen (the effect of the Bohr shift)? What is the Bohr shift?

c. Is the Bohr shift more likely to occur after exercise or after hyperventilation (rapid breathing with no exercise involved)? Explain.

3. Compare these oxygen dissociation curves for hemoglobin.

Normal adult

Normal condition

Reduced pH conditions

Reduced pH conditions

Fetal vs. Adult

a. How do the curves differ?

    i. Normal versus reduced pH?

    ii. Infant versus adult?

b. What would happen if fetal and adult hemoglobin had the same affinities for oxygen?

4. If they don't get what they want, some small children threaten to "hold their breath until they die." Very strong-willed children may turn blue, but they will not die because they will pass out and the autonomic breathing response will take over. What types of control are involved in holding your breath? Why (for what physiological reason) will you pass out if you hold your breath for too long?

Name _____   Course/Section _____

Date _____   Professor/TA _____

Activity 45.2 How is mammalian heart structure related to function?

1. Diagram and describe the path a red blood cell takes from a capillary in your big toe to your heart and back to your big toe. At each point in the pathway, indicate whether the red blood cell is most likely picking up or losing oxygen. Indicate also the relative blood pressure that part of the path is likely to have. Include all of the following terms in your diagram.

    Terms

aorta	left atrium
venules	right ventricle
veins	left ventricle
arteries	coronary arteries
arterioles	heart
capillaries	internal organs (for example, lungs, right atrium digestive tract)
left atrium	skeletal muscle

2. The degree of musculature differs in these chambers of the heart: atria, right ventricle, and left ventricle.

a. Draw the heart and its chambers, including differences in musculature.	b. Explain why the differences in musculature might exist by explaining the normal functions of each chamber.	c. Include in your description the functions of the SA node (pacemaker), the AV node, and the AV and semilunar valves.

3. While in utero, the wall between the two atria of the human fetal heart is not complete. An opening, the foramen ovale, allows blood from the two atria to mix. Normally, at birth, this hole seals over and the two atria are separated from each other. What would be the consequences to the infant if this hole did not seal over at birth?

4. One of the most common congenital defects of the cardiovascular system is called "transposition of the great arteries." In infants who have this defect, the pulmonary artery exits from the heart where the aorta should and the aorta exits where the pulmonary artery should. All other circulatory connections are normal.

   a. Diagram and describe the circulation of blood in an infant who has this congenital defect.

   b. What type of treatment would such an infant need?

Name _____ Course/Section _____

Date _____ Professor/TA _____

Activity 45.3 How are heart and lung structure and function related to metabolic rate?

1. Describe the simplest type of closed circulatory system. Include the function of each part of the system, and explain how the structure of each part is related to its function.

2. If you want to increase the rate of metabolism in an organism, you need to increase (a) the rate of food and oxygen delivery to the cells (or body) and (b) the rate of removal of waste products from the cells (or body). In the evolution of the vertebrates, what modifications of the circulatory system and/or circulatory function have occurred to increase metabolic rate?

3. Both reptiles and mammals obtain their energy via aerobic respiration:

$$glucose + 6O_2 \rightarrow 6CO_2 + 6H_2O + 36ATP$$

Both also need to be able to maintain internal homeostasis in order to survive.

a. Given this, would you expect the lungs of the lizard (an ectotherm) to have more or less alveoli per unit area than the lungs of the rat (an endotherm)? Assume the overall *SA/V* ratios of the rat and lizard are about the same. Explain the physiological reasoning behind your answer.

b. On average, would you expect the heart rate of the lizard to be higher or lower than the heart rate of the rat? Explain the physiological reasoning behind your answer.

4. Based on your knowledge of *SA/V* ratios, metabolic rates, and cardiopulmonary physiology, would you expect the following to be higher or lower for a human infant as compared to an adult? Explain the physiological reasoning behind your answers.

a. body temperature

b. heart rate

c. oxygen consumption rate/unit mass of body

Questions 5 to 7. At rest, an average human's mean arterial blood pressure (MABP) is 120/80. As a person increases his or her activity, a variety of physiological modifications must be made to support the increased activity.

For example, assume you are late for an exam and run 4 blocks to get there on time. Which of the following would occur during your 4-block run? Indicate whether the following statements are true or false. Explain your reasoning.

T/F   5.   As you run, your muscles will require more oxygen per unit time than when walking.

T/F   6.   The percentage of carbon dioxide in your blood will increase, and your blood pH will increase.

T/F   7.   Your MABP will increase because both the heart rate and stroke volume will increase.

Questions 8 to 11. Refer to the dissociation curves for two mammalian hemoglobins to answer the following questions.

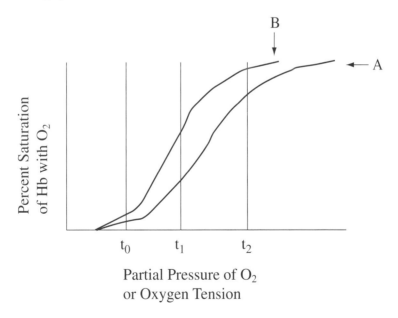

8. At oxygen tension t_0, both hemoglobins have released most of their oxygen.

9. At oxygen tension t_1, hemoglobin B is carrying less oxygen than hemoglobin A.

10. If comparison of curves A and B illustrates the Bohr effect in a single species, then curve B represents the dissociation at the higher concentration of carbon dioxide (lower pH).

11. In addition to carbon dioxide, contracting muscles cells produce a large amount of heat energy. It has been demonstrated that an increase in temperature, independent of carbon dioxide levels, encourages the blood to give up its oxygen to exercising tissues. Given this information, would you expect a or b below to be true? Explain your answer.

    a. Curve A (above) would be the lower-temperature curve and B would be the higher-temperature curve.

    b. Curve B (above) would be the lower-temperature curve and A would be the higher-temperature curve.

Activity 46.1 How do ion concentrations affect neuron function?

Much of our understanding of neuron function was based on studies of the squid's giant axon. Squid move through the water by contracting the muscles of the mantle. This movement compresses water inside the mantle and forces it out the siphon. The squid can change the direction of movement by directing the flow from the siphon either forward or backward (relative to its head or anterior end).

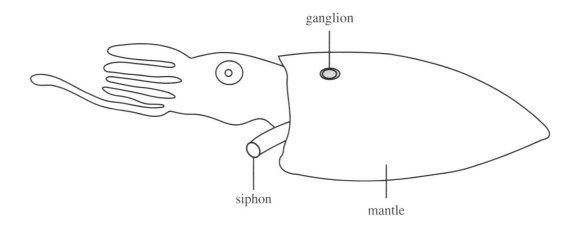

1. In which direction would the squid move if water flow from the siphon were directed toward the head end of the organism?

2. When squid are startled or in danger, they can simultaneously contract all muscles of the mantle to forcefully jet water out of the siphon and escape rapidly. Assume the mantle of the squid is 30 cm in length (about 12 inches) and the distance from the brain to the mantle opening is another 8 cm (about 3 inches). For all muscles of the mantle to contract simultaneously, a nerve signal sent from the brain must reach all parts of the mantle at the same time.

a. On the diagram of the squid above, draw three neurons. Assume all of them originate from the ganglion indicated and the ganglion is stimulated by the action of a single motor neuron from the brain. Neuron 1 innervates the mantle muscles nearest the brain. Neuron 2 innervates the muscles in the mid region of the mantle. Neuron 3 innervates the muscles at the tail end of the mantle.

b. In invertebrates, such as the squid, how must the nervous system design be modified to allow both the muscles nearest the brain and those farthest from it to contract simultaneously? Modify your drawing to indicate any necessary changes in the size or structure of the three neurons that would be required. Explain your reasoning.

3. Researchers discovered that they could remove these "giant axons" of the squid. With some skill, these axons could be maintained outside the body if held in ion solutions of the same concentrations as the squid's extracellular fluids. If stimulated with a microelectrode, these isolated axons would generate action potentials. By recording the potential difference between an electrode in the axon versus one in the fluid bathing the axon, researchers could also record the potential difference at rest and any change in potential difference that occurred as a nerve impulse, or action potential, was generated.

a. The ion concentrations the researchers recorded for the squid neuron are listed in the following table.

Ion concentrations (mM) for intra- and extracellular fluids of the squid axon		
	Intracellular (mM)	Extracellular (mM)
Potassium (K^+)	400	20
Sodium (Na^+)	50	440
Chloride (Cl^-)	40–150	560
Calcium (Ca^{2+})	1×10^{-4}	10

Some of these ion concentrations for the human are listed in the table below.

Ion concentrations (mM) for intra- and extracellular fluids of the human axon		
	Intracellular (mM)	Extracellular (mM)
Potassium (K^+)	140	5
Sodium (Na^+)	15	150

Using the Nernst equation, calculate the equilibrium potential for potassium and sodium in the human.

The Nernst equation: $E_{ion} = \dfrac{58 \text{ mV}}{z} \times \dfrac{\log [\text{ion}]_{out}}{[\text{ion}]_{in}}$ z = ion valance
(e.g., Na = +1, Cl = −1)

b. What are the equilibrium potentials for potassium and sodium for the squid?

4. In a typical neuron, which ion has the greatest influence on the membrane potential at rest? In other words, flux of which ion contributes the most to the resting membrane potential? Explain why this occurs.

5. In calculating the resting and action potentials of the axon, we generally don't worry about the concentration differences of calcium or chloride ions. Explain why this is so.

6. What effects would the following changes in extracellular K^+ concentration be likely to have on the resting membrane potential of neurons in a human?

   a. Change extracellular K^+ from 5 mM to 2 mM.

      $E_K =$

   b. Change extracellular K^+ from 5 mM to 20 mM.

      $E_K =$

Name _____   Course/Section _____

Date _____   Professor/TA _____

 ## Activity 46.2 How do neurons function to transmit information?

Working in groups of three or four, construct a dynamic (claymation-type) model of the transmission of an action potential along a neuron and then across a synapse to generate an action potential in a postsynaptic neuron.

When developing and explaining your model, be sure to include definitions or descriptions of the following terms and structures.

Neurons or parts of neurons	Ions	Gates
dendrite	K^+	voltage-gated ion channels
axon	Na^+	Na^+ gates or channels
cell body	negative organic ions	K^+ gates or channels
synaptic vesicles	Ca^{2+}	Ca^{2+} gates or channels
presynaptic neuron		
postsynaptic neuron		

Building the Model

- Using chalk on a tabletop or a marker on a large sheet of paper, draw the membranes of two neurons and the synaptic region between them. Each neuron should each be at least 4 inches across and a foot or more in length.
- Identify the axon, cell body, and dendrite(s) on your drawing.
- Make the ions and gates from modeling clay or cut-out pieces of paper. Indicate the placement of gates in the membranes and ions inside the membrane versus outside the membranes.
- Include a key for your model that indicates how ions and gates are differentiated from each other. You may use color coding for the ions and gates.
- Start the model by "initiating" an action potential in the axon hillock.
- Indicate how this action potential is propagated along the axon and how it can lead to production of an action potential in the postsynaptic neuron.
- When you have completed your model, explain it to another student or to your instructor.

Use your understanding of how action potentials are generated and propagated to answer the following questions.

1. All cells maintain an ionic (and therefore electrical) potential difference across their membranes. In most cells, this potential difference is between –50 and –100 mV. That is, the inside of the cells is more negative than the outside by 50 to 100 mV. Although all cells in the body maintain this potential difference across their membranes, only certain cells (for example, neurons) are capable of generating action potentials.

    a. How is this potential difference across the cell membrane generated?

    b. What characteristics of membranes allow cells to concentrate or exclude ions?

    c. What is it about neurons (nerve cells) that make their properties different from those of other cells? In other words, what enables nerve cells to produce action potentials?

    d. How is an action potential started and propagated?

e. Does generating an action potential require any direct or indirect energy input? If so, when and where is the energy used?

f. What happens in time and space (along the axon) once an action potential begins?

g. What factors ultimately limit the ability of the system to respond (that is, to continue to generate impulses)?

2. If an axon is stimulated in the middle of its length, nerve signals (action potentials) will move out from the point of stimulus in both directions. Normally, however, nerve signals move in only one direction along neurons. Explain.

3. Whether or not an action potential is generated in a postsynaptic neuron depends on a number of factors. What are these factors? What ultimately determines whether an action potential is generated in the postsynaptic neuron?

4. Diffusion is efficient over only very short distances. In fact, as you can see in the table, diffusion is efficient only for distances of about 1 to 100 µm.

Diffusion distance (µm)	Time required for diffusion
1	0.5 msec
10	50 msec
100	5 sec
1000 (1 mm)	8.3 min

a. How wide is a synapse?

b. If a synapse were two times as wide, what effect would it have on the transmission of nerve signals from one neuron to the next? How would this change affect the response time of an organism?

5. If you examine neuron transmission within an organism, you will discover that every action potential generated is stereotyped; for example, every action potential reaches the same maximum height and the same minimum height. In addition, the generation of action potentials is an all-or-none phenomenon. That is, once the potential difference across the membrane reaches threshold, an action potential will be generated. Given this, how does the nervous system signal differences in intensity of signal?

Activity 46.2

Name _____   Course/Section _____

Date _____   Professor/TA _____

Activity 46.3 What would happen if you modified a particular aspect of neuron function?

In the questions below, test your understanding of the various parts of the nervous system by asking yourself what would happen if (some part) was damaged. What would the system still be able to do? What would it be unable to do?

1. Some nerve gases and insect poisons work by destroying acetylcholine esterase. Acetylcholine esterase is normally present in acetylcholine synapses and acts to degrade acetylcholine. What is likely to happen to nervous transmission in insects exposed to this type of insect poison?

2. The pufferfish (fugu) contains the poison tetrodotoxin. Some shellfish produce a paralytic poison called saxotoxin. Both of these poisons block the Na^+ channels in neurons. What specific effects could these toxins have on neuron function?

3. A type of spider (the funnel-web spider) produces a toxin that blocks Ca^{2+} channels.

    a. Can a neuron exposed to this toxin fire an action potential? Explain.

    b. Can a neuron transmit a signal across the synapse using neurotransmitters? Explain.

4. You isolate a section of a squid giant axon and arrange an experiment so that you can change the solution bathing the axon. You insert an electrode into the axon and place another electrode outside the cell so that you can measure the potential across the cell membrane. With the axon bathed in normal extracellular fluid, you observe a resting potential of 70 mV and action potentials, when stimulated, that reach +55 mV.

| Ion | Concentration of each ion (mM) | | | |
| | Normal concentrations | | Experimental concentration in (a) | |
	Inside neuron	Outside neuron	Inside neuron	Outside neuron
Na^+	50	440	50	440
K^+	400	20	400	40

a. You change the solution bathing the neuron by increasing the K^+ concentration to 40 mM. What effect will this have on the neuron? For example, will it depolarize the membrane and make it easier to start an action potential? Will it hyperpolarize the membrane and make it more resistant to starting an action potential? Or will it have no effect? Explain your answer.

b. What would happen if, instead of adding more K^+ to the outside, you added more Na^+ to the fluid bathing the neuron? Explain.

Name _____   Course/Section _____

Date _____   Professor/TA _____

 Activity 46.4 How is the nervous system organized?

Neurons are single cells composed of processes called dendrites that carry information to the cell body and an axon that carries information away from the cell body. In simple terms, nerves are bundles of dendrites, or axons, or mixed axons and dendrites. Ganglia are collections of cell bodies, and brains are collections of ganglia. Processing and integration of the information occurs as a result of interaction within and between ganglia.

1. Responding to stimuli requires at a minimum:

    (a) Reception of a stimulus, which triggers
    (b) an action potential in a sensory neuron, which synapses with
    (c) the central nervous system and generates and action potential in
    (d) a motor neuron, which causes a response in
    (e) a muscle or gland.

In the space below, draw this minimum system. Be sure to name each neuron and label each of its parts.

2. The blink-of-an-eye response is one type of simple reflex response that occurs, for example, in response to water splashing up out of the sink or to a snowball aimed at your face. When this occurs, you blink, and only afterward do you "realize" why you blinked.

    a. What would you have to add to your diagram above to allow you to realize why you blinked?

    b. Why does it take you longer to realize why you blinked than it takes for the blink reaction to occur?

3. What composes the central nervous system (CNS), and how does it differ in general form and function from the peripheral nervous system (PNS)?

4. The autonomic system is subdivided into the sympathetic and parasympathetic systems. How do the general functions of each of these divisions interact to maintain homeostasis?

Autonomic nervous system divisions:	Functions:
Sympathetic	
Parasympathetic	

Activity 46.4

5. Fibromyalgia is a condition that affects about 2% of the U.S. population. Its primary symptoms are chronic pain, difficulty sleeping, and fatigue that is not relieved following sleep. Other reported symptoms include jaw clenching at night, dryness of mouth and eyes, an increased need to urinate, irritable bowels, muscle numbness and tingling, and headaches.

   a. Several researchers have proposed that fibromyalgia results from failed regulation of part of the autonomic nervous system. If this is true, failure of the regulation of which division of the autonomic nervous system could account for these symptoms? Explain your reasoning.

   b. To test their ideas, the researchers recorded the heart rates of both controls and individuals with fibromyalgia for 24-hour periods of normal activity. They observed that patients with fibromyalgia showed greatly elevated heart rates both while active and when asleep.

      Does the addition of this information change or support the hypothesis you developed in part a above?

Name _____   Course/Section_____

Date _____   Professor/TA_____

Activity 48.1 How does sarcomere structure affect muscle function?

Working in groups of three or four, use playdough or cut-out pieces of paper to construct a dynamic model of the sarcomere and demonstrate the sliding-filament model for muscle contraction.

When developing and explaining your model, be sure to include definitions or descriptions of the following terms and structures.

Terms

thick filament	sarcomere
thin filament	contracted length
actin	extended length
myosin	troponin complex
Z line	tropomyosin
H zone	Ca^{2+}
cross-bridges	

Alternatively, assign at least three people to model different parts of the sarcomere and demonstrate its action in muscle contraction. Assign a different role to each person, and assign specific actions to each role.

When you have completed your model, explain it to another student or to your instructor.

Use your understanding of the model to answer the following questions.

1. How are sarcomeres arranged within muscle fibers (and therefore within muscles)?

2. Describe the sliding-filament model of muscle action. For example, what interactions power the movement of filaments in association with each other? Include troponin complex, Ca^{2+} binding sites, tropomyosin, myosin binding site, and actin in your discussion.

3. How do your answers to questions 1 and 2 help explain why, at maximum contraction, a muscle measures 70% of its extended (uncontracted) length?

4. Review Figure 48.6 and the associated text in *Biological Science,* 5th edition. Then answer these questions.

   a. Acetylcholine, an excitatory neurotransmitter, is the neurotransmitter released into the neuromuscular junction. This release can trigger an action potential along the length of the muscle membrane. Describe the process involved.

   b. If acetylcholine always produces an excitatory response in the neuromuscular junction, how can we regulate which muscles (in the arm, for example) are contracted and which are extended at any given time? For example, how can we bend the arm only partially?

   c. Severe calcium deficiency can lead to a reduction in bone mass. It can also have serious effects on the functioning of the nervous system and on the action of muscles. Explain what roles calcium plays in the activity of muscles.

Name _____   Course/Section _____

Date _____   Professor/TA _____

Activity 48.2 What would happen if you modified particular aspects of muscle function?

Test your understanding of muscle function by asking yourself what would happen if (some part) was damaged. What would the system still be able to do? What would it be unable to do?

In each of the following experimental situations, indicate which of the proposed answers is true or correct and which is false or incorrect. If an answer is not correct, indicate why.

1. A toxin in a newly discovered bacterial strain causes irreversible inactivation of the acetylcholine receptor. If you were infected with this organism, the symptoms would include the following:

T/F a. Uncontrollable muscular contractions
    Explain:

T/F b. Muscular paralysis
    Explain:

T/F c. Loss of membrane potential of muscle fibers
    Explain:

2. In another experiment, a physiologist was able to destroy some but not all of the motor neurons innervating a specific skeletal muscle. Given what you know about motor units, what is the result of this action?

T/F   a. There is no effect, because motor neurons are redundant; multiple neurons innervate each muscle fiber as backup.

         Explain:

T/F   b. The muscle will contract faster if the physiologist destroyed inhibitory motor neurons.

         Explain:

T/F   c. The muscle will in general contract more weakly because fewer fibers are capable of contracting.

         Explain:

T/F   d. The ability of the muscle to contract will not change, because the electrical potential will spread from stimulated fibers to adjacent fibers.

         Explain:

Name _____   Course/Section _____

Date _____   Professor/TA _____

 Activity 49.1 How do hormones regulate cell functions?

This activity is designed to help you understand how hormones act within cells to produce a response.

Building the Model

Working in groups of three or four, construct a dynamic (claymation-type) model of hormone action in cells. You may use the materials provided in class or devise your own.

Step 1. Use chalk on a tabletop or blackboard to draw a eukaryotic cell. Your cell should be at least 18 inches in diameter. Be sure your drawing includes the cell membrane, the nuclear membrane, and the DNA inside the nucleus.

Assume that your eukaryotic cell responds to two different hormones:

- Hormone 1 is a protein-derived hormone. The cell responds to hormone 1 by increasing production of substance X.
- Hormone 2 is a cholesterol-derived hormone. The cell responds to hormone 2 by decreasing production of substance Y.

Step 2. Answer the following questions.

1. What structures or components do you need to add to your model to allow hormone 1 to react and increase production of substance X?

2. What structures or components do you need to add to your model to allow hormone 2 to decrease production of substance Y?

Now use playdough or cut-out pieces of paper to make your hormones, cell membrane proteins, and any other proteins you need. Indicate their placement on the membrane or cell. Include a key for your model that indicates how different hormones and proteins are designated.

Step 3. Using claymation, demonstrate how each of the two hormones is likely to produce its response.

Use the understanding you gained from your model to answer the following questions.

1. In medical applications, the type of hormone dictates the mode of administration—for example, oral versus injection.

   a. How would you need to administer hormone 1 to an organism deficient in this hormone?

   b. How would you need to administer hormone 2?

2. Hormones often act in an antagonistic fashion. That is, one hormone will initiate a certain response while another inhibits that response. Illustrate this process using insulin and glucagon as examples of antagonistic hormones.

Extend Your
Understanding   **Activity 49.2 Endocrine System: Sample Exam Questions**

1. Assume you are trying to characterize one of the hormones that are involved in the neuroendocrine regulation of milk production. You take extracts of the blood and mammary tissues of normal, lactating (milk-producing) animals and assay these extracts by injecting them into animals that have had their hypothalamus glands surgically removed; you are looking for restoration of milk production. Indicate whether each of the following findings is consistent with the hormone being a steroid, a peptide, or either type.

    a. It is found in blood.

    b. It is found in the cytoplasm of mammary tissue cells.

    c. It is found associated with receptors.

    d. It is found associated with protein complexes that contain G protein.

    e. It is found in nuclear extracts of cells.

Questions 2 to 4. Glucagon and insulin are hormones that act homeostatically to maintain glucose levels in the body. Are the following statements *true* or *false* concerning insulin and glucagon? Explain your reasoning.

T/F   2. Their production and release is stimulated by trophic hormones from the anterior pituitary.

T/F   3. They are examples of antagonistic hormones.

T/F   4. They act by stimulating storage or release of glucose from cells.

Name _____   Course/Section _____

Date _____   Professor/TA _____

  Activity 50.1 How does the production of male and female gametes differ in humans?

This activity is designed to help you understand how gamete production is controlled in mammals and particularly in humans.

In human males and females, the production of gametes and the hormones estrogen, progesterone, and testosterone is ultimately controlled by actions of the hypothalamus.

Review Chapter 50 and then, using all the terms below, diagram the control of gamete and sex hormone production first in a human male and then in a human female. Be sure to explain the role(s) of each term in your diagram.

Terms

hypothalamus   negative feedback

anterior pituitary   egg and polar bodies or sperm

LH   spermatogonium

FSH   primary and secondary spermatocytes

ovary or testes   spermatids

follicle   oogonium

corpus luteum   primary and secondary oocytes

progesterone   ootid

estrogen or testosterone

Human male:

Human female:

Name _____   Course/Section _____

Use your diagrams to answer the following questions.

1. In both males and females, the hypothalamus produces GnRH, which stimulates the pituitary to release LH and FSH. Fill in the chart.

Hormone	a. In males causes:	b. In females causes:
LH		
FSH		

2. In both males and females, the testes or ovaries produce additional hormones.

	a. Males produce:	b. Females produce:
Hormones		
Function of these hormones produced in the gonads		

3. Most birth control methods are designed to keep the egg and sperm from uniting to form a zygote. Many birth control pills or patches used by human females contain a combination of estrogen and progesterone. How do they keep the sperm from uniting with the egg? Explain the mechanism.

4. Efforts to make a male contraceptive pill (analogous to the pills used by females) have not been very successful. Given what you know about the similarities and differences in male and female gamete production, propose why this might be the case.

5. Fertilization generally occurs in the upper third of the oviduct, and development of the fetus occurs in the uterus. In some relatively rare cases, however, developing embryos have attached to the outside of the uterus and developed there for the full nine months of pregnancy.

   a. Given the anatomy of the female reproductive system, can you explain how this could happen?

   b. What modifications of normal birthing procedures (if any) would have to be made in such cases?

*Extend Your
Understanding* **Activity 50.2 Animal Reproduction: Sample exam
questions**

Questions 1 to 3. In an experiment, an adult rat's testes, including the vascular
connections, were transplanted to the wall of the abdomen. Connections of the testes to
the reproductive tract were cut. Following this rat's recovery, which of these descriptions
would be *true* or *false*? Explain your answers.

T/F   1.   The rat would have lowered sexual activity due to loss of testosterone.

T/F   2.   The rat would have normal sexual activity but be unable to produce any
ejaculate.

T/F   3.   The rate would have normal sexual activity but have no sperm in the ejaculate.

4.   A girl begins to develop breasts and pubic hair at four years of age. Given these
symptoms, her physician orders a CAT scan (imaging procedure) to look for an
endocrine tumor. Which organ would the physician most likely *not* investigate as the
cause? Explain your reasoning.

A. hypothalamus
B. pituitary
C. ovary
D. uterus

Questions 5 to 8. Assume that women can be vaccinated against the following hormones.
Each vaccine is designed to completely neutralize the target hormone's activity. Which
vaccine(s) would prevent pregnancy? Explain your answers.
A = Would prevent pregnancy; B = would not prevent pregnancy.

5.   A vaccine against LH

6.   A vaccine against CG

7.   A vaccine against estrogen

8.   A vaccine against prolactin

Activity 51.1 How does the immune system keep the body free of pathogens?

Draw a Rube Goldberg cartoon-type diagram or develop a dynamic (claymation-type) model to demonstrate how the components of the immune system interact to rid the body of a pathogen—for example, a bacterial cell or a viral particle. Be sure to explain the function of each "actor" in the system. Your diagram or model should include all the terms below.

Here is an example of a Rube Goldberg–type drawing:

Terms

bacterium or virus particle	antibody
helper T cell	antigen
cytotoxic T lymphocyte	epitope
macrophage	thymus
B cell	bone marrow
memory B cell	fever
memory T cell	self versus nonself
plasma cell	immunoglobulin
cytokines	B-cell receptor
CD4 and CD8 proteins	T-cell receptor with α and β chains
MHC molecules	

After you have completed your model or diagram, use what you have learned to answer the following questions.

1. What are pathogens? Why do we need to prevent them from colonizing our bodies? If pathogens do manage to colonize, what effects can they have?

2. What general defense mechanisms does the body use to help prevent colonization by pathogens? For example, what general defense mechanisms are involved in local inflammatory responses?

3. In specific immunity, how do B-cell responses differ from T-cell responses?

B-cell responses	T-cell responses

4. If about 10^5 genes are available in the human genome to produce proteins, how can we produce more than 10^{12} different kinds of B-cell receptors on B cells?

5. How does HIV affect the immune system?

Name _____   Course/Section _____

Date _____   Professor/TA _____

✎ Activity 52.1 What factors determine climate?

The map on the next page shows a hypothetical continent on Earth. Assume that biomes and climates on this continent are produced by the same factors that produce biomes and climates on Earth's real continents.

1. For each biome listed below, indicate its general location (e.g., latitude) and the key characteristics of its climate.

    a. Tropical wet forest

    b. Temperate grassland

    c. Subtropical desert

    d. Temperate forest

    e. Tundra

2. On the map of the hypothetical continent, indicate the location(s) of each of the biomes (listed in question 1).

   To do this, draw approximate boundary lines to delimit each biome type, and then label each delimited area with the type of biome it contains.

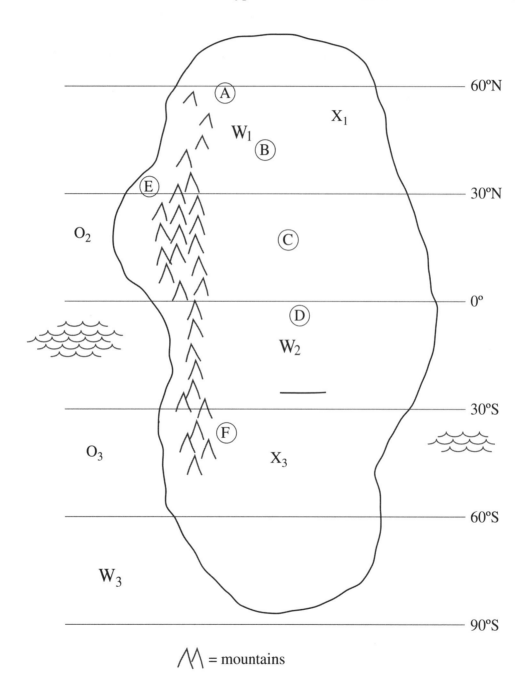

$\bigwedge$ = mountains

3. What biomes or vegetation types would most likely be found at the given points? (Assume all are at sea level or low altitudes.)

Point on the map	Most likely type of biome or vegetation
A	
B	
C	
D	
E	

4. Would the climate at point F be relatively wet or dry? Explain.

5. In the United States, the temperate forest extends from the East Coast westward for about 1200 miles (to the Mississippi River Valley). From there, the forest begins to thin out toward the west into oak savannas (or temperate woodland), and it finally gives way to open grassland (Great Plains). The grasslands extend 1000 miles westward to the foothills of the Rocky Mountains.

   a. Why does grassland replace forest west of the Mississippi River?

   b. What is the rain-shadow effect?

   c. Draw a rain shadow somewhere in the *Southern* Hemisphere of the map.

6. How are the general characteristics of plants (for example, morphology) influenced by climate? In other words, explain what effects climate has on the types of plants that grow in an area.

7. Refer to Figures 52.11–52.14 in *Biological Science*, 5th edition. In general, how is the distribution of major ecosystems or biomes related to climate? If you know the mean annual precipitation and the mean annual temperature of an area, would you be able to accurately predict the type of biome that could exist there? Explain.

8. Why isn't Earth's climate uniform? To answer this, summarize the major factors that can produce differences in climate from place to place.

9. A large asteroid hits the Earth, and alters its axis tilt from 23.5° to 10°.

   a. What effects would this have on seasons in the Northern Hemisphere?

   b. What effects would this have on biome distribution in the Northern Hemisphere?

Name _____   Course/Section _____

Date _____   Professor/TA _____

Activity 53.1 What determines behavior?

1. Some plant species (for example, many orchids) rely on a single species of insect for pollination. If the insect species dies out, so will the plant species. In parts a through c, refer only to the behavior of the insect species.

    a. What questions would you need to ask to determine proximate causation for this behavior?

    b. What questions would you need to ask to determine ultimate causation for this behavior?

    c. What kinds of experiment(s) or investigation(s) would you propose to answer at least one of the questions you posed in parts a and b?

2. Many bird species that are common in the northern United States during spring and summer fly south in the fall to overwinter and feed in Central or South America. In the spring, they return to the northern United States to breed.

   a. What questions would you need to ask to determine proximate causation for this behavior?

   b. What questions would you need to ask to determine ultimate causation for this behavior?

   c. What kinds of experiment(s) or investigation(s) would you propose to answer at least one of the questions you posed in parts a and b?

3. Many species of animals engage in complex courtship rituals. Among these species is the bower bird of Australia. Male bower birds construct elaborate structures or bowers from twigs, leaves, and moss and decorate them with colorful objects (e.g., berries and shells). The bowers and the dances the males perform are designed to attract female bower birds for mating. After mating, the females fly away to build a nest and raise the offspring. The males remain at their bowers and try to attract additional mates. These birds can live for up to 17 years. Males are territorial and build their bowers in the same location each year. In studying their behavior, researchers have noted that about 25% of the females "shop around," going from one bower to another, before deciding on a mate. The other 75% appear to go directly to a single bower to mate. These behaviors cannot be observed in captivity.

a. What questions would you need to ask to determine proximate causation for the bower's nest-building behavior?

b. What questions would you need to ask to determine the proximate causation for the female choice?

c. What questions would you need to ask to determine ultimate causation for these behaviors?

d. What kinds of experiments or investigations would you propose to answer at least one of the questions you posed in either part a or b?

Name _____   Course/Section _____

Date _____   Professor/TA _____

Activity 54.1 What methods can you use to determine population size?

1. To measure the population density of the chipmunks in a particular park, you sample several quadrants and capture 50 chipmunks. You mark each of their backs with a small dot of red paint and then release them. The next day, you capture another 50 chipmunks. Among the 50, you find 10 that are marked.

   a. Using the mark-recapture formula, below, how many chipmunks do you estimate the population contains?

   $$m_2/n_2 = n_1/N$$

   or

   $$\frac{\text{Number of recaptures in second catch}}{\text{Total number in second catch}} = \frac{\text{Number marked in first catch}}{\text{Total population } N}$$

   b. What effect would each of the following discoveries have on your estimate?

      i. You later discover that you sampled the one area of the park that was most favored by the chipmunks.

      ii. You later discover that the chipmunks were licking the marks off of each others' backs.

      iii. You later discover that the marked chipmunks are easier to see and therefore more susceptible to predation.

c. What modifications could you make to your sampling program to help ensure more accurate estimates of population size?

2. The following table provides the numbers of deaths per acre per year resulting from the operation of two different agents of mortality in populations of grasshoppers at several different density levels.

a. Fill in the mortality rates for agents A and B in this table.

Grasshopper population density (individuals/acre)	Deaths per year per acre		Mortality rate (%)	
	Agent A	Agent B	Agent A	Agent B
100	4	0		
1000	40	25		
10,000	400	500		
100,000	4000	50,000		

b. Graph the data below.

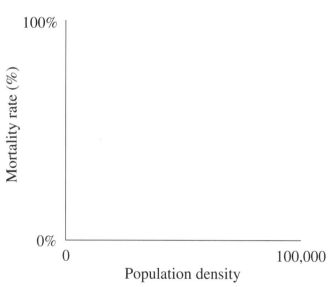

c. Which of the two agents of mortality (A or B) is operating in a density-independent manner? Explain your answer.

d. Which of the two agents of mortality (A or B) is likely to act as a factor stabilizing the size of the grasshopper population? Explain your answer.

Name _____   Course/Section _____

Date _____   Professor/TA _____

Extend Your
Understanding **Activity 54.2 Population Ecology: Salamanders**
                 and Their Parasites

A researcher has recently discovered three species of parasites that infect developing
salamanders. He suspects that one or more of these species cause fatalities during
salamander development. To test his hypothesis, the researcher sets up a series of
experiments. In each experiment, he varies the density of the salamander populations.
He infects 2% of each test population at the start of the experiment and measures the
mortality rate after four weeks. To establish a baseline mortality rate, he sets up a control
experiment that differs only in that no parasite was introduced at any density. The data
table and graph below relate the mortality rate of salamanders (caused by the three
different parasites) to the original density of the developing salamander population.

						CONTROL	CONTROL
Density A	Death Rate A	Density B	Death Rate B	Density C	Death Rate C	Density D	Death Rate D
20	0.1	20	0.82	20	0.4	20	0.1
40	0.2	40	0.74	40	0.25	40	0.096
80	0.4	80	0.76	80	0.49	80	0.12
160	0.6	160	0.8	160	0.35	160	0.15
320	0.85	320	0.76	320	0.36	320	0.092

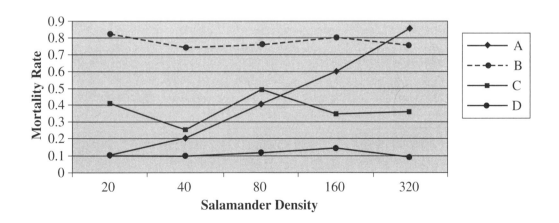

a. Given the data presented in the graph, indicate whether each of the parasite species is acting in a density-dependent or density-independent fashion. Explain your answers.

b. After looking at the data for parasites A and B, develop an argument to indicate which is more likely to cause extinction of the salamander population. Explain your reasoning.

Activity 54.3 What models can you use to calculate how quickly a population can grow?

1. In the simplest population growth model, $\Delta N/\Delta t = r_{max}N$.

    a. What does each term stand for?

Term	Stands for
ΔN	
Δt	
r_{max}	
N	

    b. What type of population growth does this equation describe?

    c. What assumptions are made to develop this equation?

2. Population growth may also be represented by the model $\Delta N/\Delta t = r_{max}N \, [(K - N)/K]$.

    a. What is K?

    b. If $N = K$, then what is $\Delta N/\Delta t$?

    c. Describe in words how $\Delta N/\Delta t$ changes from when N is very small to when N is large relative to K.

    d. What assumptions are made to develop this equation?

3. You and your friends have monitored two populations (population A and population B) of wild lupine for one entire reproductive cycle (June year 1 to June year 2). By carefully mapping, tagging, and censusing the plants throughout this period, you obtain the data listed in the chart.

Parameter	Population A	Population B
Initial number of plants	500	300
Number of new seedlings established	100	30
Number of the initial plants that died	20	100

a. Calculate the following parameters for each population:

Parameter	Population A	Population B
B (number of births)		
D (number of deaths)		
b		
d		
r_{max}		

b. Given the initial population size and assuming that the population is experiencing exponential growth at growth rate r, what will the number of plants be in each population in 5 years? (Use the initial population size as time 0, and compute to time 5.)

Population A:

Population B:

4. Using the exponential growth formula, you can determine the amount of time it will take for a population to double in size if you know r_{max} or r. The doubling time is equal to

$$\log_{10}2/\log_{10}(1 + r)$$

Alternatively, this value can be estimated by using the formula:

70 divided by the percent increase per unit time (as a whole number) = doubling time per unit time or $70/r$ = approximate doubling time.

Using either of these formulas—the exponential growth formula or the approximate doubling rate formula—calculate the following.

a. If the population of a country is growing at 2% per year, how many years will it take for the population to double?

b. If your bank account is growing at the rate of 1% per year, how many years will it take for your money to double?

5. You are studying the growth of a particular strain of bacteria. You begin with a tiny colony on a petri plate. One day later, you determine that the colony grew and exactly doubled in size. A calculation showed that if the colony continued to grow at the same (constant) rate, it would cover the entire plate in 30 days. (Assume that colony size is directly proportional to the number of individual bacteria.)

   a. What is the value of r?

   b. On what day would the bacteria cover half of the plate?

6. You collect data on birth and mortality in three populations of grasshoppers, and you calculate the following birth and death rates for these populations. Both populations are experiencing exponential growth:

	b	d
Population A	0.90	0.80
Population B	0.45	0.35
Population C	0.15	0.05

Are the following statements true or false?

T/F   a.   Population A is growing at the fastest rate

T/F   b.   Population C has the lowest death rate.

T/F   c.   Population C is growing at the slowest rate.

T/F   d.   All populations are growing at the same rate.

7. In a herd of bison, the number of calves born in 1992, 1993, and 1994 was 55, 80, and 70, respectively. In which year was the birth rate greatest?

8. A population of pigeons on the west side of town has a per capita annual growth rate of 0.07. A separate population of pigeons on the east side of town has a per capita annual growth rate of 0.10. If the populations are both growing exponentially, and both are censused the following year, in which of the populations will $\Delta N / \Delta t$ be greatest?

9. A rabbit population has the life table shown here.

Age class	Number of survivors	Number of deaths	Mortality rate	Number of offspring per reproducing pair
0–1	100	10	0.10	0
1–2	90		0.33	1.5
2–3	60	30		2.0
3–4	30	24	0.80	2.5
4–5		6	1.0	0

    a. Fill in the missing data in the table.

    b. Owing to a good food supply and a small predator population, the rabbit population is growing by leaps and bounds. The rabbits call a meeting to discuss population control measures. Two strategies are proposed:

- Delay all rabbit marriages until age class 2–3 (rabbits *never* breed until after marriage).
- Sterilize all rabbits in age class 3–4.

    Which of the proposed strategies will be more effective in slowing population growth? Explain your reasoning, and show your calculations.

Name _____   Course/Section _____

Date _____   Professor/TA _____

Extend Your
Understanding **Activity 54.4 Logistic Population Growth in Fisheries**

Suppose you have a "farm" on which you grow, harvest, and sell edible freshwater fish. The growth of the fish population is logistic. You want to manage your harvest to maintain maximum yields (that is, the maximum rate of production) from your farm over a number of years.

As a fisheries manager, you are responsible for deciding how many walleye can be harvested without destabilizing the population.

a. Below is a data table showing the walleye population in a typical pond on your fish farm over 24 weeks. Draw a graph showing how population size in the pond changes through time.

Time (weeks)	1	2	3	4	5	6	7	8	9	10	11	12
Population size	100	101	102	103	104	106	110	115	125	140	155	172
Time (weeks)	13	14	15	16	17	18	19	20	21	22	23	24
Population size	188	201	209	217	221	225	229	233	235	237	238	239

b. How large should you let the population get before you harvest? Identify the point on your graph, and explain why this is the best size for harvest.

c. Assume the carrying capacity for your pond is 250 individuals. Check your answer in part b by using the data in the chart and computing the change in the population size ($\Delta N/\Delta t$) when the population is at several different levels relative to its carrying capacity. Use $K = 250$ and $r_{max} = 0.20$.

Population size (N)	$(K - N)/K$	$\Delta N/\Delta t$
25 (low)		
50 (moderately low)		
125 (half K)		
200 (moderately high)		
250 (high)		

Name _____   Course/Section _____

Date _____   Professor/TA _____

Activity 55.1 What do you need to consider when analyzing communities of organisms?

Understanding problems in community ecology most often requires the integration of ecological principles.

For questions 1–5, analyze the situations described. Then explain which of the following ecological principles could be active in each particular situation.

Ecological principles:

coevolution

realized niche

fundamental niche

tolerance

competitive exclusion

niche differentiation or resource
    partitioning

character displacement

top-down versus bottom-up controls

trophic structure or trophic levels

keystone species

competition (inter- and intraspecific)

disturbance

1. A small clan of hyenas killed an antelope. While they were feeding on the carcass, two female lions approached, growled at the hyenas, and chased them away from the carcass.

2. Two species of closely related swallows live in England. The black swallow lives in coniferous forests, and the yellow swallow lives in deciduous forests. In Ireland, where the black swallow has never been introduced, only the yellow swallow is present; it lives in both coniferous and deciduous forests.

3. In a woodland community, three species of rodents coexist: voles, field mice, and shrews. All three species eat seeds and nuts. Each species has a preference for seeds of the most appropriate sizes for their teeth and mouths; however, all three species compete for the same kinds of nuts. An owl species also lives in this woodland community. The owl preys on all three rodent species. During one particular year, a parasite that causes pneumonia in birds is introduced into the community. This parasite dramatically reduces the owl population, which remains low for several years as a result. Following the initial reduction in the owl population, there is a dramatic increase in the population of field mice and a dramatic decrease in the populations of both voles and shrews.

4. In 1962, five mute swans escaped from captivity and began a breeding population in Chesapeake Bay. Today, there are over 4000 mute swans living in the bay. Each year they eat approximately 10.5 million pounds of aquatic grasses. These grasses provide habitat for waterfowl and crustaceans, improve water quality, decrease erosion, and increase dissolved oxygen concentrations in the bay. The birds are also aggressively territorial and have been known to trample the nests of other birds (e.g., least terns and black skimmers) and drive away native birds such as tundra swans and black ducks from feeding and roosting areas.

5. Known as the "Hawaiian woodpecker," the 'akiapola'au (aki-a-pul-a-ow) is found only in montane mesic old-growth koa/'oh'i forests, and only on the Big Island (Hawaii). It has a distinctive beak that is like a multiple-use tool.

The short, straight lower mandible is used to peck holes in the wood; the long, curved upper mandible is used to probe for insects and larvae. Males have larger beaks than females do and feed on the trunks of trees. Females feed higher on branches and twigs. 'Akiapola'au are thought to have the lowest reproduction rate for a small bird—only one chick per year, which is cared for by the parents for six months or more. The decline in their numbers appears to correspond with the introduction of rats, cats, and logging on the island.

Name _____   Course/Section _____

Date _____   Professor/TA _____

Extend Your
Understanding   **Activity 55.1 What can we learn about a community of**
                    organisms by removing one species?

A researcher collected data on an experiment she conducted on two desert islands. The islands were of similar size, climate, species composition, and richness, and they were the same distance from the mainland. Originally, the same species of snake was present on both islands (A and B). In her experiment, the researcher removed the species of snake from island A. For comparison, the species of snake was *not* removed from island B. She then recorded the number of animal species on each island over a period of 24 months. Her data are presented in the table on the next page.

a. Graph the data.

Time (months)	Number of species	
	A	B
1	36	36
2	38	37
3	35	34
4	33	35
5	31	36
6	32	38
7	29	40
8	29	36
9	26	38
10	24	39
11	20	36
12	18	37
13	18	35
14	13	34
15	11	36
16	10	38
17	9	40
18	10	38
19	10	36
20	8	38
21	9	39
22	10	36
23	9	37
24	8	35

b. Construct a hypothesis that explains the difference between the numbers of species present on island A versus island B over the 24-month period.

c. Propose an experimental design to test your hypothesis. Explain the reasoning behind your design.

d. What would you expect to find as a result of your experiment? Describe your expected results, and draw them on the graph below.

Name _____   Course/Section _____

Date _____   Professor/TA _____

Activity 55.2 What effects can a disturbance have on a community?

Disturbances rarely affect only one member of a community. More often, disturbances that have direct effects on a single organism produce a cascade of additional effects on other organisms.

For questions 1 and 2, analyze the ecological situations described. Then answer the questions that follow each situation.

1. A particularly popular island vacation site is home to many species of orchids. The primary income-generating businesses on the island are tourism and orchid sales.

   Disturbance: To make the island more attractive to visitors, a local politician suggested that the island be periodically "fogged" with insecticides.

   Response: This fogging reduced the insect numbers. It also appeared to reduce the number of birds and new growths of orchids.

   (*Hint:* Examine the ideas of limits of tolerance, coevolution, predation, and chemical competition.)

   a. Is it likely that the disturbance directly caused the response?

   b. If not, what other factors might be involved?

c. How could you test which factors produced the response? For example, what type of experiment could you set up?

d. What would you expect to find if the other factors you proposed caused the response?

2. Australia and New Zealand are home to a wide variety of marsupials (for example, kangaroos and pouched mammals). Until colonization by foreign traders and other developments, placentals (placental mammals) were not found in these areas.

Disturbance: Following colonization, the rabbit was introduced to Australia.

Response: The rabbit multiplied rapidly and ultimately became a pest species, doing considerable damage to both crops and natural plants.

(*Hint:* Examine the ideas of competition, competitive exclusion, predation, and coevolution.)

a. Is it likely that the disturbance directly caused the response?

b. If not, what other factors might be involved?

c. How could you test which factors produced the response? For example, what type of experiment could you set up?

d. What would you expect to find if the other factors you proposed caused the response?

3. In the boreal forest of Canada, wildfires are important disturbance factors. A single wildfire seldom burns the whole forest. Instead, it burns large patches or stands and leaves others untouched. Following a wildfire in a black spruce forest, there is usually a predictable regrowth of the vegetation, starting with ground lichens and small spruce seedlings. As the spruce trees grow and form a closed-crown canopy, feather mosses (bryophytes) are found in an increasing proportion on the forest floor. In some cases, the peat moss *Sphagnum* outcompetes the feather mosses and eventually dominates the ground cover. Because wildfires occur naturally about every 10 years, a forest stand can sometimes burn before that stage is reached, and the whole process repeats.

   a. Draw a curve of "Tree Biomass vs. Time" over a 100-year period:

      i. at the stand scale (the stand is a particular part of the forest that burned) and
      ii. at the landscape scale (composed of many forest stands).
         • Assume, at the stand scale, that when a stand burns, all trees in that stand die.
         • Assume, at the landscape scale, that there is one fire every 10 years, and that each fire burns a different stand.

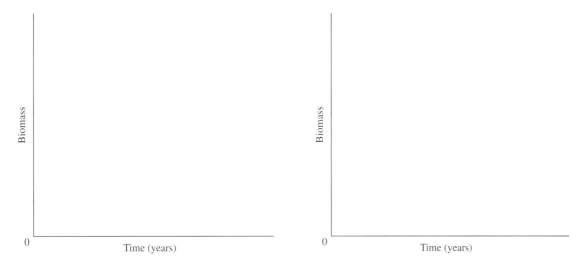

b.   Explain how (and why) the graphs differ.

Name _____   Course/Section _____

Date _____   Professor/TA _____

Activity 55.3 How can distance from the mainland and island size affect species richness?

1. Island biogeography theory attempts to explain the patterns of species richness and turnover on islands as a function of the size of the island and its distance from the mainland.

   a. What are the basic tenets of this theory (refer to Figure 55.22 of *Biological Science,* 5th edition)?

   b. In the graph below, curves A and A' represent the historic immigration and extinction rates for an island off the coast of South America. Given these data, what is the equilibrium number of species for this island?

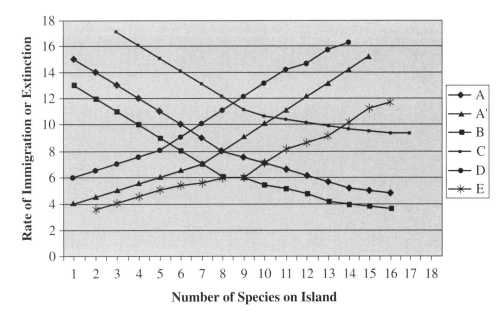

Copyright © 2014 Pearson Education, Inc.

c. If a land bridge or an isthmus forms and connects the island to the mainland, which two curves would best represent the resulting immigration and extinction rates for the island?

d. What would be the new equilibrium number of species?

Name_____   Course/Section_____

Date_____   Professor/TA_____

Extend Your
Understanding   **Activity 55.4 Reexamining the Theory of Island**
                    Biogeography

A research scientist comes back from studying a group of islands in the South Pacific. He says his data show that in some cases, the smaller islands had larger numbers of lizard species on them than did the larger islands. He suggests these data indicate that we should reexamine the whole theory of island biogeography.

a. What additional information would you like to have about the group of islands he studied?

b. Is there any way his data could be accounted for using the current theory of island biogeography? Explain your reasoning.

Activity 56.1 What limits do available solar radiation and nutrients place on carrying capacities?

1. Energy transformations in a community can be diagrammatically represented as a trophic structure.

    a. Diagram a simple pyramid of productivity for a grassland community that includes four trophic levels.

    b. At what trophic levels do humans belong in your diagram? Explain.

2. Energy is lost at every step in the trophic structure where energy is transferred—for example, from primary producers to primary consumers. It is estimated that about 1% of the solar energy striking plants is converted to chemical energy, only 50–90% of gross primary productivity (GPP) becomes net primary productivity (NPP), and less than 10% of the energy available at each tropic level transfers to the next level (as biomass).

    a. Restate the preceding information in your own words. Be sure to explain what GPP and NPP are, and indicate how they relate to the percentage of energy transfers from one trophic level to another.

b. Where does the "lost energy" go? Is it correct to say that energy is "lost" in these transfers? If not, what would be a better way of expressing this?

3. In the Hubbard Brook ecosystem, scientists estimated that solar radiation input to the forest was 1,254,000 kilocalories (kcal) per square meter. Other scientists have estimated that the Earth as a whole receives about 10^{22} joules of solar radiation each day (1 J = 0.239 cal). It has also been estimated that primary producers on Earth collectively create about 170 billion tons of organic material per year. Do the following calculations to determine how these latter two values compare.

a. Convert 10^{22} joules per day to kilocalories per year.

b. Convert 170 billion tons of organic material to kilocalories. To do this, assume all of the 170 billion tons is glucose. Also assume tons are metric tons. (*Note:* 180 grams of glucose when burned in a calorimeter gives off 686 kcal of energy.)

c. Given your answers to parts a and b, what percentage of the total incoming solar energy is captured as biomass (glucose)?

d. Measurements of photosynthetic conversion of the energy in sunlight to biomass indicate that, at best, plants can convert about 75% of the energy they absorb as sunlight into biomass. The other 25% is used to support metabolism. What could account for the apparently low efficiency of photosynthesis relative to the total incoming solar energy that you calculated in part c?

4. If we can calculate Earth's total primary productivity, we can use it to develop an estimate of the total number of humans Earth can support.

   a. How much energy in kilocalories per year would it take to support Earth's current human population if all individuals weighed about 75 kg and each required 2000 kcal per day? Assume 6×10^9 humans currently inhabit Earth.

   b. How does the estimate of the energy contained in the 170 billion tons of organic material (from part b in question 3) compare to the amount of energy required to support the current human population (from part a in question 4)?

c. Is it reasonable to assume that all of the primary productivity on Earth is available to support humans? If not, what else do you need to consider?

Questions 5 to 7. Use your understanding of energy flow through ecosystems to answer the following questions.

Two species of hawks are found at various times of the year in the United States. Species A eat mice and squirrels (both of which eat seeds). Species B eat snakes, which feed on mice and squirrels. Given what you know about trophic structures, if other characteristics of the two hawk species are similar, which of the following are likely to be true? Explain your reasoning.

A = Likely to be true.
B = Not likely to be true.

5. For a given feeding area, the number of Species B that could be supported is likely to be smaller than the number of Species A that could be supported.

6. Reduction in the available primary productivity in the feeding area would tend to have a lesser effect on Species A than on Species B.

7. If the seeds consumed by the mice and squirrels in this feeding area were contaminated with heavy metals, you would expect to find much lower concentrations of heavy metals in Species B than in Species A.

8. In addition to an energy source, primary producers require adequate mineral nutrients. You have been monitoring the net primary productivity (NPP) of a grassland area for several years. Over the years, its NPP increased initially and then leveled off. You suspect that the availability of a nutrient is limiting productivity.

   a. Design an experiment to determine whether a nutrient is limiting NPP.

   b. Given your design, what results would you expect if only one of the nutrients you test is limiting?

   c. What results would you expect if more than one of the nutrients are limiting? Would this affect your design in any way?

   d. What factor(s) other than nutrient limitation might cause NPP to level off?

Name_____   Course/Section_____

Date_____   Professor/TA_____

Activity 57.1 What factors can affect the survival of a species or community?

Making decisions to preserve communities requires an understanding and integration of many factors. Assume you work for the U.S. government and you manage a large national forest. You are told that to maintain the economy in the area, the government has agreed to allow foresters to remove half a million acres of trees from a million-acre parcel. This parcel is almost square.

You have asked your staff to do an analysis of two possible methods for implementing the plan:

- Proposal I: Split the million acres in half and allow the foresters to harvest the trees on one of the halves.
- Proposal II: Divide the million-acre tract into 50 parcels (of 10,000 acres each) and allow the foresters to harvest half of the trees in each parcel.

1. List some of the ecological advantages and disadvantages of each proposal.

Proposal I		Proposal II	
Advantages	Disadvantages	Advantages	Disadvantages

2. Given the following characteristics of various animal species, which of the forestry harvesting proposals would be more likely to ensure continued success of the hypothetical animal species described? Explain your answers.

   a. Species A is a secondary and occasionally a tertiary consumer or carnivore. Behaviorally, it roams over about 20 square miles of "home range" in search of food—for example, rabbits (herbivores) and foxes (carnivores).

   b. Species B is a small, rodentlike herbivore that exists in small numbers in the forest. Its preferred food and habitat are found along the edges of the forest and are composed primarily of herbaceous (nonwoody) annual plants that produce tender shoots in the spring and plentiful seeds later in the year.

   c. Species C is a predatory bird that feeds on small rodents and occasionally on snakes and other reptiles. It can range over large distances looking for food. It nests in hollows that form naturally in a particular species of tree. These hollows are not found in trees less than 20 years of age but are common in trees 40 years and older.

3. Conservation biologists have debated extensively which is better: many small reserves or a few large ones.

   a. What factors should be considered in making judgments about the size and location of reserves?

b. Some ecologists argue that we should be concerned about preserving the largest number of species. Others argue that we should be most concerned with saving those species judged to be of unusual importance. Develop an argument to support one of these viewpoints. Your argument should be based on our existing understanding of biology in general and ecology in particular.

4. Based on your answers to questions 1–3, which of the proposals for removing half a million acres from the national forest (question 1) would you recommend? Explain your reasoning.